LEWIS DARTNELL

Lewis Dartnell is professor of science communication at the University of Westminster. He has won several awards for his science writing, and contributes to the *Guardian*, *The Times* and *New Scientist*. He has also written for television and appeared on the BBC's *Horizon*, *Sky News*, *Wonders of the Universe*, *Stargazing Live* and *The Sky at Night*.

A tireless populariser of science, his previous books include the bestseller *The Knowledge: How to Rebuild Our World from Scratch*.

LEWIS DARTNELL

Origins

How the Earth Shaped
Human History

VINTAGE

7 9 10 8 6

Vintage
20 Vauxhall Bridge Road,
London SW1V 2SA

Vintage is part of the Penguin Random House group of companies
whose addresses can be found at global.penguinrandomhouse.com

 Penguin
Random House
UK

First published in Vintage in 2020
First published in hardback by The Bodley Head in 2019

penguin.co.uk/vintage

A CIP catalogue record for this book is available from the
British Library

ISBN 9781784705435

Printed and bound in Great Britain by Clays Ltd, Elcograf S.p.A.

Penguin Random House is committed to a sustainable future for our
business, our readers and our planet. This book is made from Forest
Stewardship Council® certified paper.

Contents

Introduction

Why is the world the way it is?

I don't mean this in a musing philosophical way – why are we all here? – but in a deep scientific sense: what are the reasons behind the major features of the world, the physical landscape of continents and oceans, mountains and deserts? And how have the terrains and activities of our planet, and beyond that our cosmic environment, affected the emergence and development of our species and the history of our societies and civilisations? In what ways has Earth itself been a leading protagonist in shaping the human story – a character with distinctive facial features, a variable mood, and prone to occasional fractious outbursts?

I want to explore how the Earth made us. Of course, each of us is literally made of the Earth, as is all life on the planet. The water in your body once flowed down the Nile, fell as monsoon rain onto India, and swirled around the Pacific. The carbon in the organic molecules of your cells was mined from the atmosphere by the plants that we eat. The salt in your sweat and tears, the calcium of your bones, and the iron in your blood all eroded out of the rocks of Earth's crust; and the sulphur of the protein molecules in your hair and muscles was spewed out by volcanoes.[1] The Earth has also provided us with the raw materials we've extracted, refined and assembled into our tools and technologies, from the roughly fashioned hand axes of the early Stone Age to today's computers and smartphones.

It was our planet's active geological forces that drove our evolution in East Africa as a uniquely intelligent, communicative and resourceful kind of ape,* while a fluctuating planetary climate enabled us to migrate around the world to become the most widely spread animal species on Earth. Other grand-scale planetary processes and events created the different landscapes and climate regions that have directed the emergence and development of civilisations throughout history. These planetary influences on the human story range from the seemingly trivial to the deeply profound. We'll see how a sustained cooling and drying in Earth's climate is the reason why most of us eat a slice of toast or a bowl of cereal for breakfast; how continental collision created the Mediterranean as a bubbling cauldron of diverse cultures; and how the contrasting climate bands within Eurasia fostered fundamentally contrasting ways of life that shaped the history of peoples across the continent for millennia.

We have become greatly concerned about humanity's impact on the natural environment. Over time our population has exploded, consuming ever more material resources and marshalling energy sources with greater and greater proficiency. *Homo sapiens* has now come to replace Nature as the dominant environmental force on Earth. Our building of cities and roads, damming of rivers, and industrial and mining activity are having a profound and lasting effect, remoulding the landscape, changing the global climate and causing

---

* Incidentally, the East African Rift was not only the evolutionary cradle and early nursery of humanity, but also the region where I spent my own childhood: attending school in Nairobi and holidays with family around the savannah, lakes and volcanoes of the Rift Valley. It's these experiences that have given me a lifelong interest in understanding our origins.

widespread extinctions. Scientists have proposed that a new geological epoch should be named to recognise this dominance of our influence over natural processes on the planet – the Anthropocene, the 'recent age of humanity'.[2] But as a species we are still inextricably linked to our planet, and the Earth's history is imprinted in our make-up, just as much as our activities have left their distinct marks on the natural world. To truly understand our own story we must examine the biography of the Earth itself – its landscape features and underlying fabric, atmospheric circulation and climate regions, plate tectonics and ancient episodes of climate change. In this book we'll explore what our environment has done to *us*.

In my previous book, The *Knowledge*,[3] I set out to solve a thought experiment: how we might reboot civilisation from scratch as quickly as possible after some kind of hypothetical apocalypse. I used the notion of the loss of all that we take for granted in our everyday lives to explore how civilisation works behind the scenes. The book was essentially an investigation of the key scientific discoveries and technological innovations that enabled us to build the modern world. What I want to do this time is broaden the perspective, to discuss not just the human ingenuity that got us to where we are today, but to follow the threads of explanation back even further. The roots of our modern world stretch far back in time, and if we trace them deeper and deeper across the changing face of the Earth, we uncover lines of causation that often take us all the way back to the birth of our planet.

Anyone who's ever chatted with children will know what I mean here. For an inquisitive six-year-old asking about how something works or why something is the way it is, your immediate answer is never satisfactory. It opens up further mysteries. A simple initial question invariably leads to a whole series of 'why?', 'but why?', 'why is that?' With an unquenchable curiosity, the child tries to get to grips with the underlying nature of the world it finds itself in. I want to explore our history in the same way, drilling downwards through more and more fundamental reasons and investigate how seemingly unrelated facets of the world in fact share a deep link.

History is chaotic, messy, random – a few years of poor rainfall lead to famine and social unrest; a volcano erupts and annihilates nearby towns; a general makes a bad decision among the sweaty clamour and gore of the battlefield and an empire is destroyed. But beyond the particular contingencies of history, if you look at our world on a broad enough scale, both in terms of time and space, reliable trends, overarching themes and dependable constants can be discerned, and the ultimate causes behind them revealed. Cultural, social, economic and political influences are of course important – but planetary processes often form a deeper layer of explanation. While our planet's make-up has not preordained everything, it can provide opportunities or constraints.

Our survey will reach over a staggering span of time. The entirety of human history has played out on an essentially static map – within but a single frame of the Earth's movie. But the world hasn't always looked like this, and although continents and oceans shift over geologically slow timescales the past faces of the Earth have greatly influenced our story. We'll look at the changing nature of the Earth and the development of life on our planet over the past few *billion* years; the evolution of humans from our ape ancestors over the last five *million* years; the increase in human capabilities and dispersal around the world over the past *hundred thousand* years; the progression of civilisation over the last *ten thousand* years; the most recent trends of commercialisation, industrialisation and globalisation of the last *millennium*; and finally how we have come to understand this wondrous origin story over the last *century*.

In the process we'll travel to the ends of history – and beyond. Historians decipher and interpret humanity's written accounts to tell the story of our earliest civilisations. Archaeologists brushing the dust off ancient artefacts and ruins can tell us about our earlier prehistory and lives as hunter-gatherers.

Palaeontologists have pieced together our evolution as a species. And to peer even further back through time we will turn to revelations from other fields of science: we will be browsing the records preserved in the layers of rocks that make up the very fabric of our planet; we will be reading the ancient inscriptions of the genetic code stored in the DNA library inside each of our cells; and we will be peeping through telescopes to survey the cosmic forces that shaped our world. The narrative threads of history and science will be intertwined throughout the book, making up the warp and weft of its fabric.

Every culture has developed its own origin story – from the Australian Aborigines' Dreamtime to the Zulus' creation myth. But modern science has built up an increasingly complete and fascinating account of how the world around us came to be, and how we took our place within it. Rather than relying solely on our imagination, we can now elucidate the chronicle of creation by using these tools of investigation. This, then, is the ultimate origin story: the tale of the whole of humanity and also that of the planet we live on.

We'll explore why the Earth has been experiencing a prolonged cooling and drying trend over the past few tens of millions of years, and how this created the plant species we came to cultivate and the herbivorous mammals we domesticated. We'll investigate how the last ice age enabled us to disperse across the globe, and why it is that humanity only came to settle down and develop agriculture in the current interglacial period. We'll look at how we have learned to extract and exploit a huge diversity of metals from the crust of the planet that have driven a succession of revolutions in tool-making and technology throughout history; and how the Earth gave us the fossil energy sources that have powered our world since the Industrial Revolution. We'll discuss the Age of Exploration in the context of the fundamental circulation systems of the Earth's atmosphere and oceans, and how seafarers came to understand wind patterns and ocean currents step by step to build transcontinental trade routes and maritime empires. We'll explore how the Earth's history has created the geostrategic concerns of today, and continues to influence modern politics – how the political map of the south-eastern US continues to be shaped by sediments from an ancient sea that existed 75 million years ago, and how voting patterns in Britain reflect the location of geological deposits dating to the Carboniferous Period 320 million years ago. It is through knowing our past that we can understand the present, and prepare ourselves to face the future.

We'll begin our ultimate origin story with the most profound question of all: what planetary processes drove the evolution of humanity?

Chapter 1

The Making of Us

We are all apes.

The human branch of the evolutionary tree, called the hominins, is part of the wider animal group of the primates.* Our closest living relatives are the chimpanzees. Genetics suggest that our divergence from the chimps was a long and drawn-out process, beginning as early as 13 million years ago, with interbreeding continuing until perhaps 7 million years ago.[1] But eventually our evolutionary histories did separate, with one side giving rise to today's common and bonobo chimpanzee, the other branching into the different hominin species, with our own kind, *Homo sapiens*, forming just one twig. If we look at our development in this way, humans didn't evolve *from* apes – we *are* still apes, in the same way that we're still mammals.

All the major transitions in the evolution of hominins took place in East Africa. This region of the world lies within the rainforest belt around the equator of the planet, on a level with the Congo, the Amazon and the tropical islands of the East Indies. By rights, therefore, East Africa ought to be densely forested too, but is instead characterised by mainly dry, savannah grasslands. While our primate ancestors were

* We'll come back to the planetary event that saw the emergence of primates as a group in Chapter 3.

tree-dwellers, surviving on fruit and leaves, something drastic happened in this region of the world, our birthplace, to transform the habitat from lush forest to arid savannah, and in turn drive our own evolutionary trajectory from tree-swinging primates to bipedal hominins hunting across the golden grasslands.

What are the planetary causes that transformed this particular region to create an environment in which smart, adaptable animals could evolve? And as we are only one of a number of similar intelligent, tool-using hominin species to have evolved in Africa, what were the ultimate reasons why *Homo sapiens* prevailed to inherit the Earth as the sole survivor of our evolutionary branch?

GLOBAL COOLING

Our planet is a restlessly active place, constantly changing its face. Fast-forwarding through deep time you'd see the continents gliding between myriad different configurations, frequently colliding and welding together only to be ripped apart again, with vast oceans opening and then shrinking and disappearing. Great chains of volcanoes pop and fizzle, the ground shivers with earthquakes, and towering mountain ranges crumple out of the ground before being ground away back to dust. The engine powering all this fervent activity is plate tectonics, and it is the ultimate cause behind our evolution.

The outer skin of the Earth, the crust, is like a brittle eggshell encasing the hotter, gooier, mantle beneath. The crustal shell is cracked, fragmented into many separate plates that rove across the face of the Earth. The continents are made up of a thicker crust of less dense rocks, while the oceanic crust is thinner but heavier and so doesn't ride as high as the continental crust. Most of the tectonic plates are made up of both continental and oceanic crust, and these rafts are constantly jostling for position with each other as they bob on top of the hot churning mantle and ride the whims of its currents.

Where two plates butt into one another, along what is known as a convergent plate boundary, something has got to give. The leading edge of one of the two plates is shunted beneath the other and is dragged down into the rock-melting heat of the mantle, triggering frequent earthquakes and feeding an arc of volcanoes. Because the rocks of the continental crust are less dense and so more buoyant, it is almost invariably the oceanic crust portion that sinks beneath the other in a plate collision. This subduction process continues until the intervening ocean has been swallowed, and the two chunks of continental crust become welded together, a great crumpled chain of mountains marking the impact line.

Divergent, or constructive, boundaries are the places where two plates are being pulled apart from each other. Hot mantle from the depths rises up into this rent, like blood welling into a gash in your arm, and solidifies to form new rocky crust. Although a new spreading rift can open up in the middle of a continent, ripping it in two, this fresh crust is dense and low-lying and so becomes flooded over with water. Constructive boundaries form new oceanic crust – the Mid-Atlantic Ridge is one prominent example of such a seafloor spreading rift.[2]

Plate tectonics is an overarching theme of the Earth we'll return to throughout the book, but for now we'll focus on how the climate change it drove over recent geological history produced the conditions for our own creation.

The past 50 million years or so have been characterised by a chilling of the global climate. This process is called the Cenozoic cooling, and it culminated 2.6 million years ago in the current period of pulsing ice ages that we'll look at in detail in the next chapter. This long-term global cooling trend has been largely driven by the continental collision of India into Eurasia and the raising of the Himalayas. The subsequent erosion of this towering ridge of rocks has scrubbed a lot of carbon dioxide out of the atmosphere, resulting in a reduction of the greenhouse effect that was previously insulating the planet (see Chapter 2), and leading to declining temperatures. In turn, the generally cooler conditions drove less evaporation from the oceans to create a less rainy, drier world.

Although this tectonic process happened some 5,000 kilometres away across the Indian Ocean, it also had a direct regional effect within the theatre of our evolution. The Himalayas and Tibetan Plateau have created a very powerful monsoon system over India and South-East Asia. But this huge atmospheric sucking effect over the Indian Ocean also drew moisture away from East Africa, reducing the rainfall it experienced. Other global tectonic events are thought to have contributed to the aridification of East Africa. Around 3–4 million years ago Australia and New Guinea drifted north, closing an ocean channel known as the Indonesian Seaway as they did so. This blockage constricted the westward flow of warm South Pacific waters, and instead colder waters from the North Pacific flowed through to the central Indian Ocean. A cooler Indian Ocean reduced evaporation which in turn meant less rainfall for East Africa.[3] But most significantly, another huge tectonic upheaval was happening in Africa itself that was to prove instrumental in the making of us.

A HOTBED OF EVOLUTION

About 30 million years ago a plume of hot mantle rose up beneath north-eastern Africa. The land mass was forced to swell upwards by about a kilometre[4] like a huge zit. The skin of continental crust over this swollen dome stretched and thinned until eventually it began to rip open right across the middle in a series of rifts. The East African Rift tore along a roughly north–south line, forming an eastern branch through what is now Ethiopia, Kenya, Tanzania and Malawi, and a western branch that cuts through Congo and then continues along its border with Tanzania.

This Earth-ripping process was more intense towards the north, tearing right through the crust to allow magma to seep through the long wound and create a new crust of basalt rock. Water then flooded into this deep rift to create the Red Sea; another rift became the Gulf of Aden. The seafloor spreading rifts tore off a chunk from the Horn of Africa to form a new tectonic plate, the Arabian. The Y-shaped meeting of the African Rift, Red Sea and Gulf of Aden is known as a triple junction and right at the centre of this intersection is a low-lying triangle of land called the Afar region, stretching across north-east Ethiopia, Djibouti and Eritrea.[5] We'll return to this important region later.

The East African Rift runs for thousands of kilometres from Ethiopia to Mozambique. As the swelling from the magma plume bulging beneath it continues, the Rift is still being pulled apart. This 'extensional tectonic' process is causing whole slabs of rock to fracture along faults and break off, with the flanks being pushed up as steep escarpments and the blocks in between subsiding to form the valley floor. Between about 5.5 and 3.7 million years ago this process created the current landscape of the Rift: a wide, deep valley half a mile above sea level and lined on both sides with mountainous ridges.[6]

One major effect of the swelling of this crustal bulge and the high ridges of the Rift was to block rainfall over much of East Africa. Moist air blowing over from the Indian Ocean is forced upwards to higher altitudes where it cools and condenses, falling as rain near the coast. This creates drier conditions further inland – a phenomenon known as a rain-shadow.[7] At the same time, the moist air from the central African rainforests is also blocked from moving eastwards by the highlands of the Rift.[8]

The upshot of all these tectonic processes – the creation of the Himalayas, the closing of the Indonesian Seaway, and in particular the uplift of the high ridges of the African Rift – was to dry out East Africa. And the formation of the Rift changed not only the climate but also the landscape, in the process transforming the ecosystems of the area. East Africa was remoulded from a uniform, flat area smothered in tropical forest, to a rugged, mountainous region with plateaus and deep valleys, its vegetation ranging from cloud forest to savannah to desert scrub.[9]

Although the great rift started to form around 30 million years ago, much of the uplift and aridification happened over the past 3–4 million years.[10] Over this time, the same period that saw our evolution, the scenery of East Africa shifted from the set of *Tarzan* to that of *The Lion King*.[11] It was this long-term drying out of East Africa, reducing and fragmenting the forest habitat and replacing it with savannah, that was one of the major factors that drove the divergence of hominins from tree-dwelling apes. The spread of dry grasslands also supported a proliferation of large herbivorous mammals, ungulate species like antelope and zebra that humans would come to hunt.

But it wasn't the only factor. Through its tectonic formation the Rift Valley became a very complex environment, with a variety of different locales in close proximity: woods and grass-lands, ridges, steep escarpments, hills, plateaus and plains, valleys, and deep freshwater lakes on the floor of the Rift.[12] This has been described as a mosaic environment, offering hominins a diversity of food sources, resources and opportunities.[13]

The widening of the Rift and the upwelling of magma was accompanied by strings of violent volcanoes spewing pumice and ash across the whole region. The East African Rift is dotted with volcanoes along its length, many of which formed in just the last few million years. Most of these lie within the Rift Valley itself, but some of the largest and oldest are growing on the edges, including Mt Kenya, Mt Elgon, and Mt Kilimanjaro, the tallest mountain in Africa.

The frequent volcanic eruptions spilled lava flows that solidified into rocky ridges cutting across the landscape. These could be traversed by nimble-footed hominins, and along with the steep scarp walls within the Rift may have provided effective natural obstacles and barriers for the animals they hunted. Early hunters were better able to predict and control the movements of their prey, constraining escape routes and directing them into a trap for the kill. These same features may also have offered vulnerable early humans a degree of protection and security from their own predators that prowled the landscape.[14] It seems that this rough and varied terrain provided hominins with the ideal environment in which to thrive. Early humans, who, like us, were relatively feeble and did not have the speed of a cheetah or the strength of a lion, learned to work together and take advantage of the lie of the land, with all its tectonic and volcanic complexity, to help them hunt.

It is active tectonics and volcanism that have created and then sustained these features of a varied and dynamic landscape over the course of our evolution. In fact, because the African Rift is such a tectonically active region, the landscape has changed greatly since the times of earliest human habitation. As the Rift has continued to widen, the areas once populated by hominins on the valley floor have now become uplifted onto the flanks of the Rift; today it is here that we find hominin fossils and archaeological evidence, completely removed from their original settings. And it is this great rift, the most substantial and long-lived region with extensional tectonics in the world today, that is believed to have been crucial to our evolution.

FROM TREES TO TOOLS

The first indisputable hominin for which we have discovered good fossil remains is *Ardipithecus ramidus*, which lived around 4.4 million years ago in forest lining the Awash river valley in Ethiopia. This species was roughly the same size as modern chimpanzees, with an equivalent-sized brain, and teeth that suggest they had an omnivorous diet. The fossilised skeletons indicate they still lived in trees and had only developed a primitive bipedality – the ability to walk upright on two feet. About 4 million years ago, the first members of the genus *Australopithecus* – the 'southern ape' – shared several traits with modern humans, such as a slender and gracile body-form (but still with more primitive skull shapes), and they were competent at walking bipedally. *Australopithecus afarensis*, for example, is well known from surviving fossils. One of these is the remarkably complete skeleton of a female who lived 3.2 million years ago in the Awash river valley, which came to be known as Lucy.*

Lucy would have stood at only about 1.1 metres, but had a spine, pelvis and leg bones very similar to those of modern humans. So while Lucy, and other members of *A. afarensis*,† still had a small, chimpanzee-sized brain, their skeleton clearly indicates a lifestyle of long-distance bipedal walking. Indeed, a bed of volcanic ash in Laetoli, Tanzania, has preserved three sets of footprints from 3.7 million years ago. These were probably created by members of *A. afarensis* and look remarkably like those you might leave in the sand during a stroll along the beach.

In human evolution, the development of bipedalism clearly came a long way before significant increases in brain size – we walked the walk before we could talk the talk. These *Australopithecus* fossils, together with those of the earlier *Ardipithecus* species, also show that bipedality didn't evolve as an adaptation to walking in open, grassy savannah environments as had been thought previously, but first emerged with hominins

---

* Named after The Beatles song 'Lucy in the Sky with Diamonds', which was played loudly in the excavation camp after her discovery in 1974.

† It's common when discussing organisms to abbreviate the genus name. So *Australopithecus afarensis* becomes *A. afarensis*. The dinosaur *Tyrannosaurus rex*, for example, is most popularly known simply as *T. rex*.

still living closely among trees in wooded areas.[15] But bipedalism certainly became an increasingly useful adaptation as the forests shrank and became more fragmented. Our early hominin ancestors were able to move between islands of woods, and then venture out into the grasslands. Bipedalism allowed them to see over the tall grass, and minimised the area of their bodies exposed to the hot sun, helping them to keep cool in the savannah heat. And the opposable thumbs that became so useful for holding and manipulating tools are also an evolutionary inheritance from our forest-dwelling primate ancestors. The hand crafted by evolution to grasp a tree branch pre-adapted us for holding the shaft of a club, an axe, a pen, and ultimately the control stick of a jet plane.

By around 2 million years ago the hominin species of the *Australopithecus* genus had all fallen extinct and our own genus, *Homo*, had emerged from them. *Homo habilis* ('handy man') was the first, with a gracile body-form similar to the earlier australopithecines and a brain only slightly larger.[16] A dramatic increase in the size of the body and brain, as well as a major shift in lifestyle, however, came with *Homo erectus*, which appeared around 2 million years ago in East Africa. Below the skull, the skeleton of *H. erectus* is very similar to that of anatomically modern humans, including adaptations for long-distance running and a shoulder design that would have allowed the throwing of projectiles. They are also thought to have exhibited other traits shared with us, like long childhoods of slower development and advanced social behaviour.

H. erectus was probably also the first hominin to live as a hunter-gatherer and to control fire – not just for warmth but possibly also for cooking their food.[17] They may even have used rafts to travel over large bodies of water.[18] By 1.8 million years ago *H. erectus* had spread across Africa and then became the first hominin to leave the continent and disperse through Eurasia, probably in several independent waves of migration.[19] This species persisted for almost 2 million years. By contrast, anatomically modern humans have only been around for a tenth of that time – and at the moment we'd be lucky to survive the next 10,000 years, let alone 2 million.

H. erectus gave rise to *Homo heidelbergensis* around 800,000 years ago, which by 250,000 years ago had developed into *Homo neanderthalensis* (the Neanderthals) in Europe and the Denisovan hominin in Asia. The first anatomically modern human, *Homo sapiens*, emerged in East Africa between 300,000 and 200,000 years ago.

Over the course of human evolution, hominins became increasingly bipedal, and then more efficient long-distance runners,[20] with changes to the skeleton including an S-shaped spine, a bowl-like pelvis and longer legs to support this upright posture and mode of locomotion. Body hair became reduced, except on the scalp. The shape of the head also transformed, producing a smaller snout with a more pronounced chin, and a more bowl-shaped brain case.[21] Indeed, the major difference between the earlier *Australopithecus* genus and our *Homo* lineage was this increase in brain size. Throughout their 2 million years of evolution the australopithecines' brain size was strikingly constant at around 450 cm³, roughly equivalent to that of a modern chimpanzee. But *H. habilis* had a brain a third larger, at about 600 cm³, and brain size doubled again from *H. habilis* through *H. erectus* to *H. heidelbergensis*. By 600,000 years ago, *H. heidelbergensis* had a brain roughly the same size as that of modern humans, and three times larger than that of australopithecines.[22]

As well as increasing brain size, another defining feature of the hominins was how we applied our intelligence to tool-making. The earliest widespread stone tools – known as Oldowan technology – date back to about 2.6 million years ago, and were used by the later *Australopithecus* species as well as *H. habilis* and *H. erectus*. Rounded cobbles from a river were used for cracking open bones or nuts against another, flat, anvil stone. Sharpened edges were created by chipping off flakes, and this shaped stone was then used for cutting and scraping meat from a kill, or for wood-working.*

A revolution in Stone Age technology came when *H. erectus* inherited Oldowan tools and refined them into the Acheulean industry 1.7 million years ago. Acheulean tools are more carefully worked by knocking off smaller and smaller flakes to create more symmetrical and thinner, pear-shaped hand-axes. They have represented the dominant technology for the vast majority of human history. A later transformation produced the Mousterian technology, used by Neanderthals and anatomically modern humans through the Ice Age. Here, the core stone was carefully prepared and trimmed by knapping around the edge, before a final, large flake was skilfully knocked off. It was the removed flake rather than the shaped core stone that was the goal: a thin, pointed shard was perfect as a knife or could be used as a spear point or arrowhead.[23]

These stone tools, as well as wooden spear shafts, enabled hominins to become fearsomely effective hunters without needing to develop large teeth or claws on their own bodies like other predators. We exploited sticks and stones as artificial teeth and claws to hunt for food or to defend ourselves, all whilst being able to keep a safe distance from prey and predators to minimise the risk of injury.

These developments in body-form and lifestyle enhanced each other. More efficient running and sophisticated cognitive abilities, coupled with tool use and control of fire, enabled more effective hunting and a diet with an ever greater proportion of meat for powering a larger brain. This in turn enabled us to

---

* Stone Age tools have been found made from materials like quartzite, chert, volcanic obsidian, and flint. These rock types are mainly composed of silica: silicon dioxide. Silica has offered the base material for transformative technologies throughout our history as a species, from stone tools, to glass, to the high-purity silicon wafers of modern computer microchips. In this way, the East African Rift, for over 2 million years the centre of the cutting-edge technology (if you'll excuse the pun) of stone tool manufacture, was the original Silicon Valley.

develop more complex social interaction and cooperation, cultural learning and problem solving, and perhaps most significantly, language.[24]

THE CLIMATE PENDULUM

Many of these landmark transitions in our evolution are preserved within the Afar region – the triangular depression that as we saw sits right at the intersection of the tectonic triple junction – at the northern, and oldest, end of the Rift. The first hominin fossils, those of *Ardipithecus ramidus*, were discovered in the Awash river valley that runs north-east from the Ethiopian plateau towards Djibouti, flowing right through the middle of the Afar triangle. This same river valley preserved the 3.2 million-year-old remains of Lucy – indeed, her entire species, *Australopithecus afarensis*, was named for this region. And the oldest known Oldowan tools were found in Gona, Ethiopia, which also lies within the Afar triangle. But the whole length of the East African Rift Valley has been a hotbed of hominin evolution.

The drying climate and the rift system with its mosaic of varied features, including volcanic ridges and fault scarps, were clearly instrumental in providing the environmental conditions to drive our evolution. But while this complex, tectonic landscape may have provided opportunities for roaming hominins, it doesn't explain sufficiently how such incredible versatility and intelligence emerged in the first place. The answer is thought to come down to a particular quirk of the extensional tectonics of the great Rift Valley, and how it interacts with fluctuations in the climate.

As we have seen, the world has been getting generally cooler and drier for the past 50 million years or so, and the tectonic uplift and formation of the Rift Valley has meant that East Africa in particular dried out and lost its former forests. But within this global cooling and drying trend, the climate became very unstable and swung back and forth dramatically. As we will discover in more detail in the next chapter, around 2.6

million years ago the Earth slid into the current epoch of the ice ages, with its alternating glacial and interglacial phases driven by rhythmical shifts in Earth's orbit and tilt known as the Milankovitch cycles. East Africa was too far from the poles to encounter the advancing ice sheets themselves, but this doesn't mean it wasn't greatly affected by these cosmic cycles. In particular, the periodic stretching of Earth's orbit around the Sun into a more elongated egg shape – known as the eccentricity cycle – has produced periods of highly variable climate in East Africa. During each of these phases of extreme variability, the climate oscillates back and forth between very arid and wetter conditions, with the faster beat of the precession cycle of Earth's tilted axis, which we'll come back to.[25]

Still, these cosmic periodicities and the swings in climate they drive operate over thousands and thousands of years. If we want to understand human evolution, the mystery is that processes which have had the biggest influence on East Africa – such as the overall drying effect from tectonic uplift and rifting within the region, or climate rhythms like the precession of Earth's axis – operate on an exceedingly slow timescale compared to the lifespan of an animal. Yet intelligence, and the extremely versatile behaviour it allows, is an adaptation similar to the use of a multi-tool Swiss army knife, helping an individual cope with diverse challenges as the environment varies significantly within its lifetime. Environmental changes over a much longer timescale can be met by evolution adapting the body or physiology of a species over the generations (such as the camel adapting to constantly arid conditions). Intelligence on the other hand is the evolutionary solution to the problem of an environment that shifts faster than natural selection can adapt the body. So for there to have been a strong evolutionary pressure driving hominins to ever more flexible and intelligent behaviour, something must have been affecting our ancestors over very short timescales.

What might have been special about the circumstances in East Africa that drove evolution towards highly intelligent hominins such as ourselves? The answer that has been emerging in recent years comes down once again to the peculiar tectonic environment of the region. As we have seen, East Africa was bulging upwards with the magma plume rising beneath and this stretched the crust until it fractured and faulted. The geography of the Great African Rift is therefore characterised by a flat valley floor where great chunks of crust have sunk down, and which is lined on both sides by mountainous ridges. In particular, from about 3 million years ago numerous large, isolated basins formed on the valley floor that could fill with lakes if the climatic conditions were wet enough.[26] These deep lakes are important because they provided hominins with a more reliable source of water through the dry seasons each year than that supplied by streams.[27]

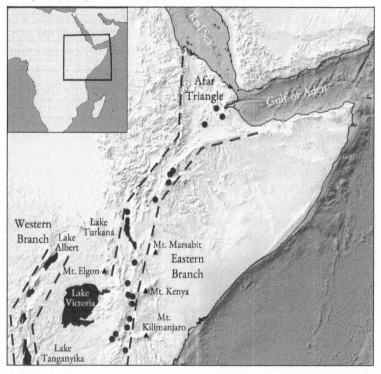

The East African rift valley system, with major lakes and amplifier lake basins shown.

But many of them were also ephemeral: they appeared and disappeared over time with the shifting climate.

The landscape of the tectonic rift creates a sharp contrast in the conditions between the high ground and the bottom of the valley. Rain falls over the tall rift walls and volcanic peaks, where it then flows into the lakes dotting the valley floor, a much hotter environment with high rates of evaporation. This means that many of the lakes in the Rift Valley are exceedingly sensitive to the balance between precipitation and evaporation, and even a slight shift in climate causes their water levels to respond very considerably and rapidly – far more so than other lakes around the world and even elsewhere in Africa.[28] As small changes in the regional climate cause very large changes in the levels of these vital bodies of water, they are known as 'amplifier lakes' – they act like a hi-fi amplifier intensifying a weak signal. And it is these peculiar amplifier lakes that are thought to provide the key link between the long-term trends of tectonics creating the rift valley and the Earth's climate swings and the rapid fluctuations of habitats that directly, and dramatically, affected our evolution.

Two particular aspects of our planet's cosmic circumstances are important here: the stretching of Earth's orbit around the Sun (eccentricity) and the gyration of the Earth's axis (precession). Every time Earth's orbit was pulled into a more elongated shape (maximum eccentricity) the climate in East Africa became very unstable. During each of these phases of climatic variability, whenever the precession cycle cast a little more solar warming onto the Northern Hemisphere, more rain fell onto the walls of the Rift Valley. The amplifier lakes appeared and enlarged, their shores lined with woodland. And conversely, during the opposite phase of the precession cycle the rift received less rainfall and the lakes diminished or disappeared altogether. The Rift Valley then returned to an extremely arid state with minimal foliage.[29] So overall, the environment in East Africa over the last few million years has largely been very dry, but this general state was interspersed with highly variable periods when the climate swung rapidly back and forth between being much wetter and then very arid again.

These variable phases occurred every 800,000 years or so, and during those periods the amplifier lakes flickered in and out like a loose lightbulb – each swing causing a considerable shift in the availability of water, vegetation and food, which had a profound influence on our ancestors. The rapidly fluctuating conditions favoured the survival of hominins who were versatile and adaptive, and so drove the evolution of larger brains and greater intelligence.[30]

The three most recent periods of such extreme climatic variability occurred 2.7–2.5, 1.9–1.7, and 1.1–0.9 million years ago.[31] Looking at the fossil record, scientists have made a fascinating discovery. The timing of when new hominin species emerged – often associated with an increase in brain size – or fell extinct again, tends to coincide with these periods of fluctuating wet–dry conditions. For instance, one of the most important episodes in human evolution occurred in the variable period between 1.9 and 1.7 million years ago, a phase when five of the seven major lake basins in the rift repeatedly filled and emptied. It was during this time that the number of different hominin species reached its peak, including the emergence of *H. erectus* with its dramatic increase in brain size. Overall, of the fifteen hominin species we know of, twelve first appeared during these three variable phases.[32] What's more, the development and spread of the different stages of tool technologies that we discussed earlier – Oldowan, Acheulean, Mousterian – also correspond with the eccentricity periods of extreme climate variability.[33]

And not only did the variable periods determine our evolution, they are also thought to have been the force driving several hominin species to migrate out of their birthplace and into Eurasia. We'll explore in detail in the next chapter how our species *Homo sapiens* were able to disperse around the entire globe, but the conditions propelling hominins out of Africa in the first place again lie with the climate fluctuations in the Great Rift.

During each wet phase the filling of the large amplifier lakes and the extra availability of water and food would cause a population boom, while at the same time limiting the amount of space available for habitation along the tree-lined rift shoulders. This would have squeezed hominins along the tube of the Rift Valley and eventually pushed them out of East Africa with each wet pulse of the precessional cycle, like a climate pump. Moister conditions would also have allowed hominin migrants to move north along the Nile tributaries and across the greener corridors of the Sinai Peninsula and Levant region to spill into Eurasia.[34] *Homo erectus* left Africa during the variable climate phase around 1.8 million years ago, eventually spreading as far as China. In Europe, *H. erectus* evolved into the Neanderthals, while the *H. erectus* population that remained in East Africa eventually gave rise to anatomically modern humans 300,000–200,000 years ago.

Our own species dispersed out of Africa around 60,000 years ago, as we'll see in the next chapter. We encountered the descendants of previous waves of hominin migrants – Neanderthals and Denisovans – as we moved through Europe and Asia. But both of these had died out by around 40,000 years ago, and only anatomically modern humans remained. From a peak in the diversity of different hominin species in Africa about 2 million years ago,[35] through our interactions (and interbreeding) with closely related human species as we moved through Eurasia, *Homo sapiens* became a lonely species. We are today the sole survivor of our genus, and indeed of the entire hominin tree.

This in itself is a curiosity. We know from the extensive archaeological evidence that the Neanderthals were themselves an extremely adaptable and intelligent species. They crafted stone tools and hunted with spears, controlled fire, and may have decorated their bodies and even buried their dead. They were also physically stronger than us *Homo sapiens*. And yet almost as soon as we arrived in Europe the Neanderthals disappeared. They may have succumbed to the punishing climatic conditions in the depths of the Ice Age (although the uncanny coincidence with our arrival would seem to discount that explanation), or perhaps anatomically modern humans violently clashed with these pre-existing Europeans and we slaughtered them into oblivion. But the most likely explanation is that we simply outcompeted them for resources in the shared environment. Modern humans are thought to have had a much better capability with language and thus social coordination and innovation, as well as more advanced tool-making abilities. And despite having dispersed from tropical Africa more recently, we could craft sewing needles and so were able to make warmer, body-hugging clothing as the Ice Age dipped into particularly bitter spells.[36]

Humans prevailed over the Neanderthals with our brains not our brawn, and subsequently came to dominate the world. And the reason for this is probably the fact that our ancestors had a longer evolutionary history in the extreme fluctuating climate of East Africa, forcing them to developing greater versatility and intelligence than the Neanderthals. We had spent longer adapting to the wet–dry variability of the Rift Valley, and that also made us better able to cope in the different climates we encountered around the rest of the world, including the Ice Age climes of the Northern Hemisphere.[37]

All in all, the human animal was forged by a peculiar combination of planetary processes all coming together in East Africa over the past few million years. It wasn't just that the region had dried out as Earth's crust bulged with the magma plume rising beneath, shifting from the relatively flat, forested habitat of our primate ancestors into arid savannah. The entire landscape had become transformed into a rugged terrain cut through with steep fault scarps and ridges of solidified volcanic lava: this was a world fragmented into a complex mosaic of different habitats that continued to change with time. In particular, the extensional tectonics of East Africa ripped open the Rift Valley to create a particular geography of tall walls collecting rainfall and a hot valley floor. Cosmic cycles in Earth's orbit and spin axis periodically filled basins on the rift floor with amplifier lakes that responded rapidly to even modest climate fluctuations to create a powerful evolutionary pressure on all life in this region.

These unique circumstances of our hominin homelands drove the development of adaptable and versatile species. Our ancestors came to rely more and more on their intelligence and on working together in social groups. This diverse land-scape, varying greatly across both space and time, was the cradle of hominin evolution, and out of it emerged a naked and chatty ape smart enough to come to understand its own origins. The hallmarks of *Homo sapiens* – our intelligence, language, tool use, social learning and cooperative behaviour, which would allow us to develop agriculture, live in cities and build civilisations – are consequences of this extreme climatic variability, itself produced by the special circumstances of the Rift Valley. Like all species, we are a product of our environment. We are a species of apes born of the climate change and tectonics within East Africa.[38]

WE ARE THE CHILDREN OF PLATE TECTONICS

Plate tectonics did not just create the diverse and dynamic environment of East Africa in which we evolved as a species; they were also to be a factor that defined where humanity embarked on building our early civilisations.

If you look at a map of the tectonic plate boundaries grinding against each other and superimpose the locations of the world's major ancient civilisations, an astonishingly close relationship reveals itself: most are located very close to plate margins. Considering the amount of land available for habitation on Earth, this is a startling correlation, and is very unlikely to have come about by chance. Early civilisations seem to have chosen to snuggle up close to tectonic fractures, millennia before scientists identified their existence. There must be something about the plate boundaries that made them so favourable for the establishment of ancient cultures, despite the dangers of earth-quakes, tsunamis and volcanoes posed by these fractures in the Earth's crust.

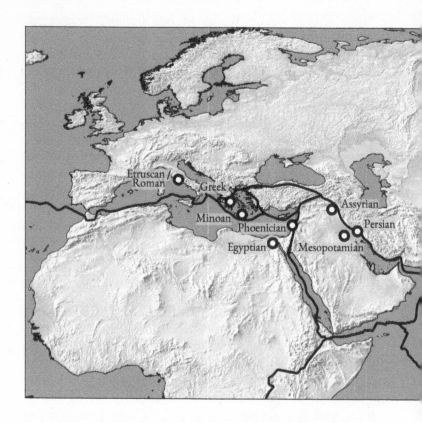

In the Indus Valley, the Harappan civilisation emerged around 3200 BC as one of the three earliest in the world (alongside those in Mesopotamia and Egypt),[39] in a depressed trough running along the foot of the Himalayas. The collision between tectonic plates creases up ranges of high mountains – such as the Himalayas, created by the crashing of India into Eurasia – but the immense weight of the mountain range also flexes the crust alongside it downwards to create a low-lying subsiding basin. The Indus and Ganges rivers flowing off the Himalayas run through this foreland basin, where they deposit sediment eroded from the mountains to produce very fertile soils for early agriculture. You could say that the Harappan civilisation

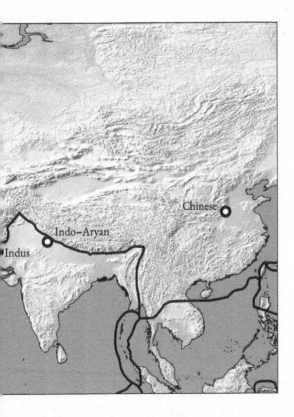

Major early
civilisations and
their proximity to
plate boundaries.

was born of the continental collision between India and the Eurasian plate.

In Mesopotamia, the Tigris and Euphrates rivers also flow along a subsiding foreland basin, pushed down by the Zagros mountains that formed as the Arabian plate was subducted beneath the Eurasian (shown on page 71).[40] Mesopotamian soils were therefore similarly enriched with sediment eroded out of this mountain range.[41] The Assyrian and Persian civilisations both arose right on top of this junction between the Arabian and the Eurasian plate.

The Minoans, Greeks, Etruscans and Romans all also developed very close to plate boundaries within the complex tectonic environment of the Mediterranean basin. Within Mesoamerica, the Mayan civilisation emerged from around 2000 BC and spread across much of south-eastern Mexico, Guatemala and Belize, with major cities built among the mountains raised by the subduction of the Cocos plate beneath the North American and Caribbean plates. And the later Aztec culture flourished close to the same convergent plate boundary, with its earthquakes and volcanoes like Popocatepetl, the 'Smoking Mountain', sacred to the Aztecs.*

And it is not just depressed basins at the feet of mountain ranges raised by continental collisions, like Mesopotamia, that hold rich arable land. Volcanoes also produce fertile agricultural soil. They arise in a broad line 100 kilometres or so away from the subduction line, as the swallowed plate sinks deeper into the hot interior and melts to release rising bubbles of magma to feed eruptions on the surface above. Civilisations in the Mediterranean, such as the Greek, Etruscan and Roman, arose in areas of rich volcanic soil in the band where the African plate is being subducted under the smaller plates making up the Mediterranean region.[42]

Tectonic stresses also hold open fractures in rocks or push up blocks of crust in what is known as a thrust fault, which often create water springs. The long line of linked mountains along southern Eurasia, crumpled up by the collision of the African, Arabian and Indian plates, happens to coincide with the arid band across the Earth's surface. This includes the

---

* The two main exceptions to this pattern of early civilisations arising on tectonic boundaries were those in Egypt and China. But Egyptian civilisation was supported by the regular flooding of the Nile, depositing fertile sediment eroded out of its headlands in the mountains surrounding the tectonic rift valley in Ethiopia and Rwanda. And Chinese civilisation began in the plain of the Yellow River in the north before spreading south into the valley of the Yangtze river, both of which flow down from the Tibetan plateau thrust up by the continental collision of India and Eurasia. So although not located along a plate margin, both Egyptian and Chinese civilisations still owe their agriculture – and wealth – to recent tectonic features.

Arabian and Great Indian deserts, and is created by the dry, descending portion of circulation in the atmosphere (which we'll come back to in Chapter 8). Here these thrust faults frequently lie at the junction between low-lying barren deserts and high-rising inhospitable mountains or plateaus, and so trade routes often pass along these geological boundaries. Towns dotted along the way accommodate the travelling merchants, supported by the water springs at the foot of the mountains.[43] But while tectonic movements can provide water sources in otherwise arid environments, these settlements are also vulnerable to destructive earthquakes with each new slip of the crust.[44]

In 1994 the small desert village of Sefidabeh in south-eastern Iran was utterly destroyed by an earthquake. The curious thing was that Sefidabeh is exceedingly remote: one of the few stops on a long trade route to the Indian Ocean, it's the only settlement for 100 kilometres in any direction. And yet the earthquake seemed to target the village with uncanny precision. It turns out that Sefidabeh had been built right on top of a thrust fault lying far underground. The fault was so deep that it had created no obvious signs of its existence on the surface, such as a tell-tale scarp, and so hadn't been previously identified by geologists. In hindsight, the only sign was an unremarkable, gently-folded ridge running alongside the town, that had slowly been built up over hundreds of thousand years of earthquake movements. The settlement had grown here because this continual tectonic up-thrusting maintained springs at the base of the ridge – the only water source for miles around. The tectonic fault had created the conditions allowing life in the desert, but it also had the potential to kill.[45]

The sources of water provided by these thrust faults have been used for thousands of years, and explain the location of many ancient settlements on tectonic boundaries. They are becoming an increasing cause for concern in the modern world, however. The capital of Iran, Tehran, began as a cluster of small towns on a major trade route at the base of the Alborz mountain range. The city grew rapidly from the 1950s and today is densely populated with a permanent population of over 8 million residents, rising to over 10 million during working hours.[46] But the small trading towns originally occupying this site through the centuries were repeatedly damaged or levelled outright by the jerk of earthquakes as this thrust fault shifts to relieve mounting tectonic stress. The city of Tabriz, further along the mountain chain to the north-east of Tehran, was devastated by earthquakes in 1721 and 1780, each killing more than 40,000 people at a time when the population of any city was only a tiny fraction of what it is today. If, or indeed when, another large earthquake jolts on this thrust fault, the effects on Tehran could be devastating. People have settled at such thrust faults for millennia, drawn by the water supply they create and the trade routes running along the landscape boundary, and the large modern cities that have developed here are now particularly vulnerable from this geological heritage.[47]

We are the children of plate tectonics. Some of the largest cities in the world today rest on tectonic faults, and indeed many of the earliest civilisations in history emerged along the boundaries of the plates that make up Earth's crust. And more fundamentally, tectonic processes in East Africa were critical for the evolution of hominins and the forging of our particularly intelligent and adaptable species. Let's turn now to the peculiar period of our planet's history that enabled humanity to migrate from our birthplace in the Great Rift Valley and come to dominate the entire globe.

Chapter 2

Continental Drifters

We are currently living in something of a peculiar geological age. It is a time that is distinguished by a single, dominant feature: ice. This may sound surprising, given our current concerns over global warming. That average temperatures have been rising since the Industrial Revolution, and particularly rapidly over the past sixty years, is undeniably true. But this recent jump caused by human activity is occurring within the general time frame of the long-term glaciation of the Quaternary Period. About 2.6 million years ago, at the start of the latest geological period, the Earth slid into a new climatic regime, characterised by the pulse of recurring ice ages. These conditions have had profound effects on the world we find today, and how we took our place in it.

At present we are living through an interglacial period, with relatively warm conditions, shrunken ice caps and consequently higher sea levels. But the average conditions over the past 2.6 million years have been much icier than today. We are perhaps familiar from museum exhibits and TV documentaries with how the world looked in the last ice age – a time when great ice sheets expanded over much of the northern hemisphere, woolly mammoths strode across the tundra-like landscape preyed upon by sabre-toothed tigers, and fur-clad Palaeolithic humans hunted with stone-tipped spears.

Yet this was only the latest phase of glaciation in our recent planetary history. There have been between forty and fifty ice ages over the past 2.6 million years,[1] and they've been getting progressively longer and colder over time. In fact, the Quaternary is an exceptionally unstable time for the planet's climate,[2] which has been see-sawing between bitter ice ages and warmer interglacial intervals, driving the periodic expansion and contraction of huge ice sheets. The freeze-ups last on average 80,000 years, the shorter respites between ice ages only around 15,000 years.[3] Each interglacial period, such as the current Holocene Epoch we entered 11,700 years ago, is no more than a brief thermal intermission before the climate plunges back into another frosty episode. We'll see later why our planet has entered this erratic climatic phase, but let's look first at the conditions of the last ice age.

CHILLY TIMES

This began about 117,000 years ago, and lasted around 100,000 years until the beginning of the current Holocene interglacial.[4] At its peak, between 25,000 and 22,000 years ago, immense ice sheets up to 4 kilometres thick[5] extended from the north to smother northern Europe and America. Another smaller ice sheet expanded across Siberia, and great glaciers spread down from mountain ranges such as the Alps, Andes and Himalayas, as well as the rugged backbone of New Zealand.

These expansive ice sheets and glaciers locked up huge amounts of water, and the sea levels around the world dropped by up to 120 metres, exposing much of the continental shelves around the margins of the great land masses as dry ground. The North American, Greenland and Scandinavian ice sheets spread all the way to the lip of these continental shelves, and the seas around them would have been covered with floating ice layers.[6]

As well as being punishingly cold close to the ice sheets, reduced evaporation from the frigid seas would have made the world much drier. Howling winds drove fierce dust storms across the arid plains.[7] Much of Europe and North America's landscape would have been tundra-like, with the underlying soil frozen all year round (permafrost), and dry, grassy steppes stretching as far as the eye could see further south. Many of the trees that grow across Europe today survived only in isolated refuges around the Mediterranean. Twenty thousand years ago, the dense forests and woodlands of today's Central Europe would have instead resembled present-day Northern Siberia.[8]

With the end of each ice age, the oceans rose again and flooded the continental shelves. The returning interglacial climate saw the ecosystems around the world slowly spread back towards the poles, following the ameliorating conditions behind the retreating ice sheets. Migrations are common within the animal world – birds flying south for the winter or the great herds of wildebeest surging like a tide across the Serengeti – but forests also migrate. Of course, individual trees cannot uproot and move, but as the climate gets warmer, seeds and saplings survive a little further north each year, and over time the forest genuinely marches (like the prophecy in *Macbeth*). After the last ice age, tree species in Europe and Asia are estimated to have migrated north at an average rate of over 100 metres every year.[9] Animals followed them – herbivores feeding directly on plants and the predators in turn tracking them. Recurring ice ages have forced the movement of plant and animal life to sweep north and south like a living tide.

Ice ages vary in their intensity, and interglacial periods are also not all alike.[10] The last interglacial, which occurred about 130,000–115,000 years ago, was generally warmer than our current one. Temperatures were up at least 2 °C from today, sea levels around 5 metres higher, and the sort of animals you would normally associate with Africa were tramping thr⌐ Europe. When construction workers were digging i⌐ Square in the centre of London in the late 1⌐ ered the remains of a range of large a⌐ hippopotamus and elephants, as well as lio⌐ this previous interglacial period.[11] Standing

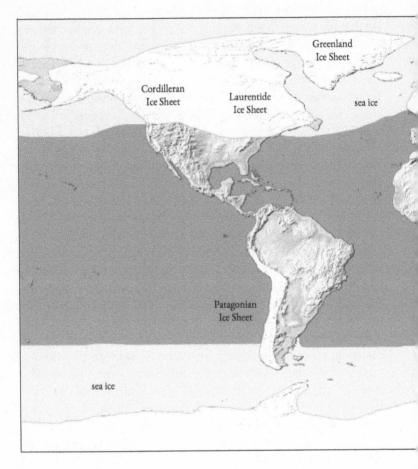

Ice Age Earth, showing the major continental ice sheets, and sea levels 120 metres lower than today.

Nelson's Column today, tourists are eager to snap selfies with the bronze lion statues sitting on guard at the corners. How many of them realise that during the last interglacial they would ʾve had to keep an eye out for the real deal?

ʾut despite the brief warmer periods that allowed these animals ʾread, the Quaternary is essentially one long ice age; even

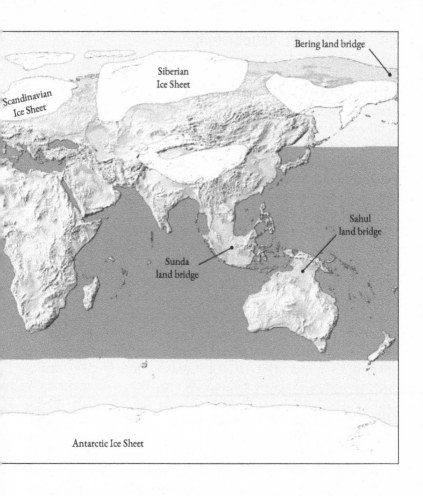

during the interglacial periods thick ice caps still smother the poles. Let's turn now to what has been going on with the Earth over recent planetary history to create such a cold, fluctuating climate. It turns out that the recurring pattern of these ice ages has cosmic causes: it can be explained by shifts in the tilt of the Earth relative to the Sun and in its orbital path.

CELESTIAL CLOCKWORK

If the Earth rotated perfectly upright there would be no seasons. It is the tilt of the planet's axis which means that for half the year the Northern Hemisphere receives more warmth than its southern counterpart as it leans towards the Sun – which appears high in the sky for its rays to shine more directly onto the surface – to create summer. The situation is reversed six months later to create the northern winter and, correspondingly, southern summer. The Earth also does not move around the Sun in a perfect circle: its orbital path is elongated slightly into an egg shape known as an ellipse. At one point in its year-long orbit the Earth is slightly nearer the Sun, and six months later it's slightly further away.*

To make things more complicated, these features of our world and its orbit also change over time, nudged by the gravitational effects of the other planets in the solar system (especially the giant Jupiter). There are three significant ways in which the Earth's cosmic circumstances vary, resulting in the set of celestial cycles which I briefly introduced in the last chapter. First, our orbit varies between a more circular and a more elongated shape over a roughly 100,000-year 'eccentricity' cycle. Second, over about 41,000 years the tilt of the Earth relative to the Sun sways back and forth between 22.2° and 24.5°, tipping the poles towards or away from the Sun. This tilt has a strong effect on the intensity of the seasons, and so even a tiny change in the angle means that the Arctic receives slightly more or less warmth in summer. And the third and shortest cycle is the 26,000 years over which the axis of the planet rolls round in a circle like a wobbling spinning top, a process known as precession. Precession changes the times of year when the Northern and Southern hemispheres are tilted towards the sun, and thus the timing of the seasons (it's also

---

* Currently, summer in the Northern Hemisphere actually falls when the Earth is furthest from the sun in its elliptical orbit.

called precession of the equinoxes). At the moment, the North Pole just so happens to be pointing towards the star called Polaris – which makes it very useful for navigators as we'll see in Chapter 8 – but in about 12,000 years' time the Earth's spin axis will have rolled around to point towards a new north star, Vega, and summer in the Northern Hemisphere will fall in what we currently call December.

So the stretch, tilt and wobble of the Earth and its orbit all have effects on the planet's climate, and they vary cyclically over time. These periodic variations are the Milankovitch cycles that I mentioned briefly in the last chapter, named after the Serbian scientist who first worked out how these cosmic periodicities change the climate on Earth. Milankovitch cycles don't on the whole reduce the total amount of sunlight warming the surface of the Earth over a year's orbit: they change the

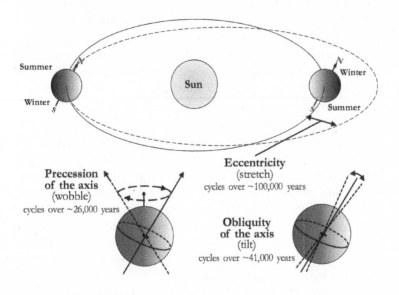

The Milankovitch cycles: variations in Earth's orbit and axis that affect our climate.

distribution of the sun's heat between the Northern and Southern hemispheres, and therefore the intensity of the seasons.

Contrary to what you might intuitively think, the key driver for triggering ice ages is not how chilly the Arctic gets in winter, but how cool the summer is. With a period of colder summers in the north, each winter's new snowfall is not completely melted away again and so can build up year on year.[12] A cooler Arctic summer also often means a warmer winter, and this too can favour the build-up of ice sheets: greater evaporation from warmer seas will deliver more snowfall. The eccentricity of Earth's orbit in particular acts as an amplifier on the effects of the direction of Earth's axis as it precesses round in a loop. For example, whenever these two cycles fall into step with each other, so that the point in our orbit when the North Pole tilts towards the Sun happens to coincide with Earth's furthest point in its elliptical circuit, then the arctic will receive unusually cool summers. And as a result, the winter's ice growth won't completely melt away and will start to accumulate. The planet begins to slide into another ice age.

The Earth remains stuck in this whitened state, reflecting away much of the sun's heat, until the Milankovitch rhythms cycle back to deliver more warmth to the north and the ice sheets thaw and retreat again.[13] The thawing at the end of each glacial period is always much quicker than the freezing at the beginning. As the Milankovitch cycles act to warm the Northern Hemisphere again, the ocean releases more carbon dioxide and more water vapour, both of which are greenhouse gases and so amplify the warming. The rising sea levels also undermine the edges of the ice sheets, and as these melt away, a greater surface area of land and sea is exposed that absorbs more sunlight than bright white ice.[14] Hence the rhythm of the ice ages is marked by a slow descent into frozen conditions, followed by rapid deglaciation.

From the start of this icehouse period about 2.6 million years ago the pulse of ice ages followed the 41,000-year beat of the Milankovitch cycle of the Earth's tilt, but for reasons that aren't yet clear about 1 million years ago this transitioned into slower but more extreme swings following the 100,000-year eccentricity cycle of Earth's stretching orbit.[15] The ice ages fell into synch with the beat of a different drum[16] – a drum that beats slower but louder. Each ice age became more intense and lasted longer: major ice sheets from the North Pole were able to advance right down the Eurasian and North American land masses, and didn't completely melt away again during interglacial warm periods.[17] (The Antarctic ice cap also waxes and wanes, although to a much lesser extent.[18])

In this sense, then, astrologers are right – just not in the way they think. The motion of the other planets through the heavens won't determine your mood or luck, but their gravitational effects on our world do influence something far more profound: the climate of the Earth itself. Understanding the celestial clockwork regulating the tick of these ice ages over the past few million years is pretty straightforward. But the subtle effects of the Milankovitch cycles can only trigger repeated swings between ice age and interglacial phases if the world's climate is already unstably perched right on the brink of glaciation. So the bigger question is: what caused these icehouse conditions in the first place?

FROM HOTHOUSE TO ICEHOUSE

At the moment, the Earth is in something of a weird period of its lifetime. For around 80–90 per cent[19] of its existence our planet has been significantly hotter than it is today; periods with ice caps at the poles are in fact something of a rarity. Over the last 3 billion years there have been perhaps only six eras with significant ice on the planet.[20] Yet over the past 55 million years, the Earth has experienced a continued chilling and the global climate has shifted from hothouse to icehouse. This is known as the Cenozoic cooling, after the geological era during which it occurred.

The layer cake of different rocks beneath our feet enables geologists to divide the long history of our planet into different eras, periods and epochs, often by referring to the kinds of fossils found within them, like chapters and paragraphs within the book of time. The current era dominated by mammals and angiosperm plants – we will return to our planet's fauna and flora in Chapter 3 – is called the Cenozoic (meaning 'new life') and began 66 million years ago with the mass extinction of species that ended the Mesozoic ('middle life') Era characterised by the dinosaurs. The most recent period within the Cenozoic is the Quaternary, defined by the fluctuating climate of glaciations and inter-glacial phases that we have just encountered. Slicing up time even more finely, the latest epoch of the Quaternary is the Holocene: our current interglacial that holds the entire history of human civilisation.

At the end of the Cretaceous Period, just before the dinosaur-killing mass extinction 66 million years ago, the world was hot and humid with lush forests growing even in polar regions. Sea levels were perhaps as much as 300 metres higher than today and submerged half the continental area on the planet – only 18 per cent of the Earth would have been dry land back then.[21] This warm phase continued for the next 10 million years, peaking with the Palaeocene–Eocene Thermal Maximum (we'll explore its significance in Chapter 3) 55.5 million years ago, before the global climate began a sustained cooling. About 35 million years ago the first permanent ice sheets appeared on Antarctica,[22] 20–15 million years ago ice sheets began forming over Greenland, and by the beginning of the Quaternary the cooling had passed the threshold for the North Pole's ice cap to begin expanding. We entered the current phase of pulsing ice ages.[23]

It seems the Earth has been committed to a concerted effort towards cooling down. What grand-scale planetary processes have been conspiring to drive this global chilling?

Gases like carbon dioxide and methane, as well as water vapour, in the atmosphere act like the panes of glass in a

Era	Period	Epoch	Millions of years ago	
Cenozoic	Quaternary	Holocene	0.0117	← Current interglacial
		Pleistocene	2.588	← Onset of Current Ice Age cycles
	Neogene	Pliocene	5.333	← Origin of hominins
		Miocene	23.03	← Spread of grass ecosystems
	Palaeogene	Oligocene	33.9	
		Eocene	56	Palaeocene–Eocene Thermal Maximum Origin of Primates and Ungulates
		Palaeocene	66	← End–Cretaceous Mass Extinction
Mesozoic	Cretaceous	Upper	100.5	▮ Major period of oil formation
		Lower	145	
	Jurassic	Upper	163.5	
		Middle	174.1	
		Lower	201.3	
	Triassic	Upper	235	
		Middle	247.2	
		Lower	252.6	← End–Permian Mass Extinction
Palaeozoic	Permian	Lopingian	259.9	
		Guadalupian	273.3	
		Cisuralian	298.9	
	Carboniferous	Pennsylvanian	323.2	▮ Major period of coal formation
		Mississippian	358.9	
	Devonian	Upper	382.7	
		Middle	393.3	
		Lower	419.2	
	Silurian	Pridoli	423	
		Ludlow	427.4	
		Wenlock	433.4	
		Llandovery	443.4	
	Ordovician	Upper	458.4	
		Middle	470	← First plants on land
		Lower	485.4	
	Cambrian	Furongian	497	
		Series 3	509	
		Series 2	521	
		Terreneuvian	541	← Origin of animals

Pangea supercontinent

Divisions of Earth's geological history

greenhouse: they allow the short-wavelength visible sunlight to shine right through and heat the Earth, but block the longer-wavelength infrared light given off by the warm planet surface. The effect of these greenhouse gases is to trap heat energy from escaping back into space, and so insulate the planet, leading to higher temperatures. Any mechanism that reduces the amount of these greenhouse gases in the air will therefore drive a global cooling.

Fifty-five million years ago, as we saw in the last chapter, the dance of the continents drove India to begin to crash into Eurasia and thrust up the huge Himalayas. Ever since, this towering mountain range has been vigorously eroded by high-altitude glaciers and rain. The minerals of the rocks react with the carbon dioxide dissolved in rainwater, which then flows in rivers to the ocean where it is used by marine life to build their calcium carbonate shells. When these creatures die, their shells drift to the seafloor and become buried. Thus the Himalayas are being gradually disassembled grain by grain and in the process carbon dioxide is locked away from the atmosphere. While this is a powerful mechanism for effectively sucking CO_2 out of the air, it still took around 20 million years to reduce the high levels of this greenhouse gas from the Cretaceous Period to below the threshold where the world became cool enough for ice to start forming at the poles.[24]

While the young Himalayas eroded, continental drift carried Antarctica to its current position over the South Pole, and Australia and South America rode north. This isolated Antarctica and opened up an unobstructed sea path right around the pole, a great oceanic moat completely surrounding the southern continent. A strong ocean current circling Antarctica became established, and this blocked warm ocean currents from the equator reaching Antarctic shores, keeping the continent chilled. The first permanent ice cap began to form on Antarctica about 35 million years ago.[25]

Plate tectonics also rearranged the other continents to shove most of the land mass into the Northern Hemisphere, while the southern half of the world became mostly open ocean (a feature we'll return to when we come to look at the powerful Roaring Forties winds in Chapter 8). For the past 30 million years or so, 68 per cent of the northern hemisphere has been continents, with only a third of the Earth's land south of the equator.[26]

This yin-and-yang division of the world – a land-dominated Northern Hemisphere and an oceanic southern half – amplifies the effects of seasonal variations in warmth from the sun. Land cools down far more quickly in winter than the turbulent ocean water, and is much better able to support growing, thick ice sheets. Yet while it is true in general that there is more land mass in the Northern Hemisphere, the pole in the Southern Hemisphere happens to have a continent – Antarctica – currently sitting right over it, whereas the North Pole is sea. This explains why the South Pole became smothered in an ice cap much earlier than its northern counterpart. At the North Pole, where ice melts more easily in the ocean, it wasn't until 2.6 million years ago that the climate became cool enough for ice no longer to melt away each summer and to accumulate year on year.

The final geological factor that created the icehouse conditions of today was the formation of the Isthmus of Panama. This narrow thread of land joining North and South America was also the result of continental collision, the plate subduction first producing a string of volcanic islands and then lifting the seafloor up above the waves. The closure of the connection between the Pacific and Atlantic oceans occurred 2.8 million years ago[27] and deflected the equatorial current to the north, strengthening the Gulf Stream that delivers warm water to the landmasses around the North Atlantic. While this current of warm water may have slightly delayed glaciation in the north, overall the extra moisture in the air from evaporation produced greater snowfall in winter and so encouraged the growth of ice sheets in the Northern Hemisphere.[28]

As the ice caps formed, first at the South and then at the North Pole, they contributed to further cooling as their bright white surface reflected more sunshine back into space – a snowball effect that scientists term a feedback loop. And as the seas became cooler they could hold more dissolved carbon dioxide from the air, further pulling down atmospheric levels and reducing the warming greenhouse effect.[29]

The effects of mountain-building and subsequent erosion removing atmospheric carbon dioxide, plate tectonics isolating Antarctica on the South Pole and forming the Panama Isthmus to alter ocean circulation patterns, continental drift shunting most of the rest of the land masses into one hemisphere – all these factors combined to propel us into icehouse conditions. The cooling of our planet to the stage where large ice sheets formed in the north 2.6 million years ago was a critical threshold and the entire climate tipped into an unstable state. Now, whenever the Milankovitch cycles acted to cool the North Pole slightly, the ice cap expanded over Europe, Asia and North America, and these large northern continents were able to support thick ice sheets. Even a small increase in the expanse of white ice would reflect more of the sunlight away, causing further cooling, and so set in motion a runaway process that saw the ice sheets expand ever further, and locking up more water out of the oceans leading to a drop in sea levels.

This sustained chilling trend of the world over the past 55 million years of the Cenozoic Era has had a profound influence on the planet, and our own evolution. As we saw in the last chapter, the shift to cooler, drier conditions shrank the forests of East Africa and replaced them with grasslands, driving the development of hominins. And the rapid fluctuations of the Rift Valley's amplifier lakes, which drove us to become a highly versatile and intelligent species, were caused by the precession rhythm of the Milankovitch cycles.

Beginning around 100,000 years ago a planetary alignment was sliding into place. The tilt of the Earth's axis causing summer in the Northern Hemisphere started to coincide with the time when the planet was furthest from the sun in its elliptical orbit, meaning that northern summers became ever cooler. The ice from each winter didn't melt away again but accumulated. The northern ice sheets began to grow and expand southwards as the Earth slid into another ice age.

Let's look now at how this most recent ice age, and the consequent drop in global sea levels, also provided a crucial opportunity for us to spread around the world. We are all children of Africa, but we didn't remain in our cradle.

EXODUS

Roughly 60,000 years ago, our ancestors began to disperse out of Africa. It's difficult to know exactly which routes we took around the planet, or the precise timing of when we first reached new areas, because the fossil record is very patchy and it is often hard to tell from the archaeological evidence exactly which branch of hominins they were left by. So most of our understanding of humanity's expansion comes from studying the genetics of indigenous populations living around the world today. By analysing the DNA, and being able to estimate the rate at which mutations accumulate in the genetic code, we can work out how long ago different populations diverged from each other. Mapping this genetic variation around the globe allows us to work out when humans first arrived in different regions, and so enables us to follow the ancient migration paths.

Two main kinds of DNA have been most useful in this detective work. Inside each of our cells are tiny structures called mitochondria, which run the biochemical reactions for providing energy. These mitochondria are the powerhouses of the cell, and they contain their own little loop of DNA. When you were conceived, you inherited the mitochondria from your mother's egg cell, but none from your father's sperm: the mitochondrial DNA passes continuously down the maternal line, from mother to daughter. Analysing the genetics of mitochondrial DNA, and calculating the time it took for different populations to split from each other, allows us to backtrack to where they converge – that particular woman in the deep past who happened to be the ancestral mother of all people alive today. This most recent matrilineal common ancestor has been dubbed Mitochondrial Eve, and she lived in Africa around 150,000 years ago. If instead we look at the DNA held on the Y chromosome that is only passed from father to son we can backtrack to the most recent male ancestor, nicknamed Y-chromosome Adam. Dating the root of this genetic tree is more uncertain, but the male common ancestor is believed to have lived between 200,000 and 150,000 years ago.

This doesn't mean that there was only one woman and one man alive at the time, or indeed that the female and male most recent common ancestors ever met each other – they lived at different times and in different places. In fact, it would be a staggering coincidence if the female mitochondrial line happened to date back to the same time as the male Y-chromosome line. (The biblical nicknames are misleading in this sense.) The only significance of Mitochondrial Eve (and analogously Y-chromosome Adam) is that she just so happened to give birth to daughters who themselves had daughters, and so on down the line to all people alive today; by chance the other lineages in the family tree died out or had no female children.

The most surprising result to come out of these global genetic studies is that the human species is exceedingly uniform. Despite superficial regional differences in hair and skin colour, or skull shape, the genetic diversity among the 7.5 billion humans living in the world today is astonishingly low.[30] In fact, there's more genetic diversity between two groups of chimpanzees living on opposite banks of a river in Central Africa than there is between humans on opposite sides of the world.[31] Human genetic diversity is greatest within Africa, however, so even if we had never discovered any fossilised bones or early archaeological evidence and had only the DNA of modern humans to go by, it would still be clear that we all originated in Africa and spread from this birthplace. Moreover, the genetic studies suggest that humanity around the world today descends from a single exodus event out of Africa, rather than multiple waves of migration, and probably from no more than a few thousand original migrants.[32]

Modern humans, *Homo sapiens*, first entered the Arabian Peninsula during a regional climatic shift to wetter conditions and a greening of the area,[33] either by walking north across the Sinai peninsula, or by taking a more southerly route by raft across the Bab-el-Mandeb strait.[34] As our ancestors began to spread into Eurasia we encountered other species of hominins that had already left Africa much earlier. Modern humans underwent a small degree of interbreeding in the Middle East with Neanderthals so that we picked up a trace of their DNA and then carried it with us as we populated the rest of the world[35] – it makes up around 2 per cent of the genetic code of non-Africans today.[36] The fact that modern East Asians appear to have more Neanderthal DNA than Europeans would suggest that humans mixed with Neanderthals on at least one other occasion as we migrated eastwards through Eurasia.

More interbreeding seems to have occurred with another mysterious, extinct hominin species known as the Denisovans when we moved through Central Asia. We know of the Denisovans only from a few teeth and the fragments of a finger and a toe bone that were discovered in a cave in the Altai mountains on the border between Siberia and Mongolia, and the DNA analysis reveals that they were probably a sister species of the Neanderthals.[37] Between 4 and 6 per cent of the DNA of modern people in Melanesia and Oceania derives from the Denisovans, and they also made a small contribution to the genetic code of Native American populations.[38] It's incredible to think that an entire human species, living alongside our own just a few tens of thousands of years ago, is known to us only by a some bone fragments and the trace of DNA they left imprinted in our genome. An even earlier hominin species, *Homo erectus*, had left Africa almost 2 million years ago[39] and reached as far as China and Indonesia, but had already fallen extinct by the time humanity spread across Asia. No indigenous peoples remaining in Africa carry DNA from either the Neanderthals or Denisovans.

As the first human migrants reached each new area their population grew and their descendants continued the dispersal. The region now covered by Iraq and Iran acted as a major dispersal hub, with migration streams heading up into Europe, across the rest of Asia, and into Australia and the Americas.[40] It seems likely that humans first headed east, following the southern margin of Eurasia to India and South East Asia;[41] an early offshoot from this path took humans into Europe around 45,000 years ago.[42] The eastward migration split into two routes either side of the Himalayas, like a river flowing around a rock, with one path heading north across Siberia and eventually into the Americas, and a second taking a southerly route across South East Asia towards Australia. The spread through southern Asia seems to have been relatively quick, possibly due to the similarity in climate with our ancestral home in sub-Saharan Africa, and we reached South East Asia and China around 50,000–45,000 years ago.[43]

From the Indochinese Peninsula, we crossed into New Guinea and Australia about 40,000 years ago.[44] With global ocean levels over 100 metres lower than today owing to ice-age conditions, the shallow seas around Indonesia were exposed as dry land. The Indonesian archipelago became part of an extension of South East Asia known as Sundaland, and Australia, New Guinea and Tasmania were all united as a single landmass called Sahul. These two lands faced each other across a narrow sea studded with chains of islands, which helped our migration into this south-eastern corner of the world.[45]

The slow wave of dispersal eventually reached the north-eastern tip of Eurasia, and it is here that the ice-age conditions proved most crucial for human migration: they gave us our route of entry into the Americas.

Today, the coasts of Russia and the US are separated by the 80-kilometre wide Bering Strait, with the two Diomede Islands sitting right in the middle of that sea channel.* During the last

---

* As Sarah Palin famously said in 2008, you can actually see Russia from Alaska. Incidentally, Russia owns the western island, Big Diomede, and the US Little Diomede. And because the international date line passes between them, these two tiny islands just a few kilometres away from each other are in time zones a whole day apart.

ice age, with the dropping sea levels, the land of Siberia and Alaska would have extended to reach towards each other, like the outstretched fingers of Adam and God painted by Michelangelo on the ceiling of the Sistine Chapel, until eventually they touched and the two vast continents of Eurasia and the Americas were joined. This land corridor would have widened until, at the glacial maximum around 25,000 years ago, it stretched up to a thousand kilometres north to south.

Although clear of ice sheets, the Bering land bridge would still have been a decidedly harsh environment: cold and dry, with dunes of silt that had been eroded by the glaciers and blown in the wind. The land bridge was little more than an Arctic wasteland, but dotted with enough hardy vegetation to support animals – woolly mammoths, ground sloths and steppe bison, as well as the sabre-toothed tigers that preyed on them.

Humans made it across this land bridge into America sometime after 20,000 years ago.[46] But other animals had already made the crossing in the opposite direction into Eurasia during an earlier ice age, some of which would become crucial to civilisations through history. Both the camel and horse had evolved in North America and crossed into Eurasia along the Bering land bridge, subsequently dying out in their birthplace. (We'll return to the significance of this in Chapter 7.)

After walking across the land bridge into Alaska, humanity worked its way down through the Americas as the ice sheets receded. At the time, two huge expanses of ice – the Cordilleran and Laurentide ice sheets – covered most of Canada and large areas of the northern USA. At its maximum extent, the Laurentide ice sheet was larger than the entire Antarctic ice cover today, with an immense dome up to 4 kilometres thick over Hudson Bay.[47] To get round these ice sheets to head south, the migrants may have travelled down the western coastline, or perhaps passed between the two along an ice-free corridor.[48] But once safely past the ice sheets in North America, humanity rapidly spread across the continent as the Ice Age subsided. They crossed the Panama Isthmus into South America around

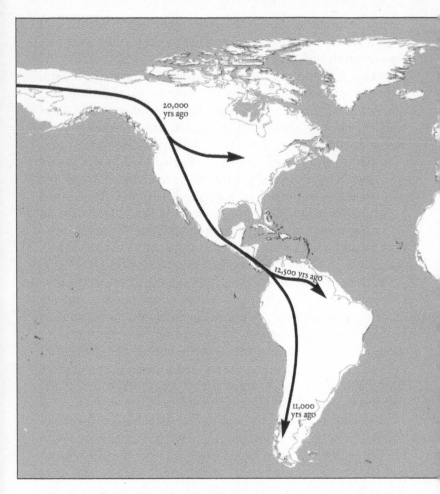

The Ice Age world with the migration paths of *H. sapiens* and the approximate ranges of Neanderthals and Denisovians.

12,500 years ago, and had reached the very southern tip of the continent within another millennium. Humanity had encompassed the globe.

Thus the Ice Age and its lowering of the global sea levels enabled the peopling of the Americas. As they had moved

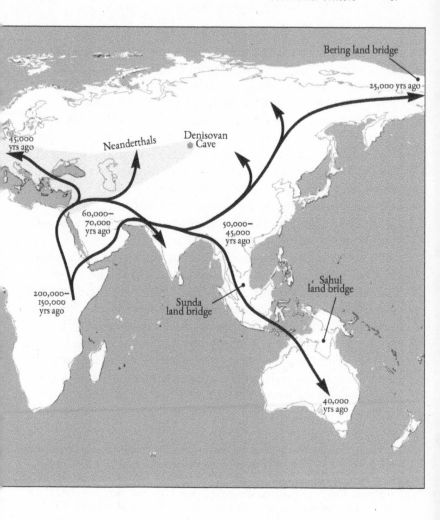

through Europe and Asia our ancestors had made contact with Neanderthals and Denisovans, but here in America they encountered no previous peoples. After they crossed the Bering land bridge into the new world, humanity was walking where no hominin species had ever trodden before.[49]

Then, around 11,000 years ago, as the world warmed again after the last glacial maximum and sea levels rose, the Bering land bridge once again disappeared beneath the waves. The connection between Alaska and Siberia was severed, and the Eastern and Western hemispheres were cut off from each other. Lasting contact was not made again between the peoples of the Old World and the New for another 16,000 years, until Columbus set foot on the Caribbean islands in 1492. Genetically similar, but living in different landscapes with access to different plants and animals, these two isolated populations of humanity formed civilisations independently from each other but remarkably similar in their domestication of crops and livestock and the development of agriculture.*

I may have given the impression that our spread was a rapid, or even directed, migration to all corners of the globe – as if our ancestors mindfully turned their backs on their original homelands in Africa and boldly strode towards the horizon, perhaps furrowing their brow in an expression of fortitude, systematically filling in all the nooks and crannies in the outline of the continents. But these movements are more accurately described as dispersals, with groups of hunter-gatherers ranging widely across the landscape with very low population densities, slowly moving with the seasons and over the years with

---

* The detective work involved in trying to trace the spread of humanity around the globe carries a lot of uncertainty in the timings and exact routes taken, with frequent disagreement between the genetic, fossil and archaeological evidence. I've presented the consensus view here, but there are claims for much earlier arrivals of humanity in China, Australia or North America. One recent controversial study, for example, argues that an unidentified hominin species reached California during the preceding ice age, 130,000 years ago.[50] What does seem likely, however, is that the exodus of modern humans out of Africa around 60,000 years ago, which gave rise to all people around the world today, was not the first. Fossilised remains in caves in Israel and stone tools found in the Arabian Peninsula[51] suggest earlier migrations around 100,000 years ago, but these apparently reached a dead end and didn't go on to populate the rest of the world.[52] It's almost as if early sparks of humanity blew out of Africa but didn't catch.

changing local climates, roving to avoid the cold and drought and to seek warmer, moister, more favourable conditions for finding food.[53] Over the generations we drifted further and further afield. Humanity's spread from the Arabian Peninsula along the southern Eurasian coast to China, for example, occurred at an average rate of less than half a kilometre a year.

Ultimately, however, humanity inherited the Earth. Our cousin hominin species – the Neanderthals and Denisovans – slid into extinction. As we discussed in the last chapter, it seems likely that they were simply outcompeted by humans rather than hunted and killed, or else succumbed to the harsh conditions as the Ice Age reached its peak. The last Neanderthals disappeared between 40,000 and 24,000 years ago, and we became the sole surviving human species on Earth. Within 50,000 years of leaving Africa we had colonised every continent apart from Antarctica to become the most widespread animal species on the planet. Our mastery of fire and skill in making clothes and wielding tools allowed us, a bunch of savannah apes, to inhabit every climate zone from the tropics to the tundra. We moved out of the environment that made us, and learned to create our own artificial habitats of huts and farms, villages and cities.*

That this global expansion took place during the punishingly cold climate of the last ice age is perhaps surprising, but it is in fact these very icehouse conditions that enabled us to accomplish this. The growth of the northern ice sheets drew so much water out of the oceans that the dropping sea levels exposed

---

* The Neanderthals weren't the only species to have been apparently greatly affected by modern humans appearing in their environment. The dispersal of humans into new geographical regions had a profound impact on local ecosystems around the world, and in particular on large animals, known as megafauna. By roughly 12,000 years ago, around a third of the large-bodied mammals in Eurasia, and about two-thirds in North America, had fallen extinct. The most likely cause is the arrival of highly-skilled human hunters, to which these large herbivores had not been previously exposed. The only continent that kept its complement of large animals was Africa, where for millions of years the megafauna had adapted to hominins while they slowly improved their hunting abilities.[54]

great areas of the continental shelves. It is the Ice Age that allowed us to simply walk across dry land to Indonesia, cross the narrow sea into Australia, and crucially make our way along the Bering land bridge into the Americas. Lower sea levels also meant that there was a much greater area of land to live on – an additional 25 million square kilometres, roughly the equivalent of present-day North America.[55]

But alongside providing the conditions that enabled humanity to spread across the globe, past ice ages have had other far-reaching implications for the moulding of the landscape we inhabit and the course of our history.

RAMIFICATIONS

You may know that the 'crinkly edges' of Norway made up of innumerable U-shaped fjords were carved by advancing glaciers during the ice ages, as were the lochs of Scotland. And even if glaciation was much less prominent in the Southern Hemisphere, if you look at a map of Chile you'll see identical fjord features along the Pacific coastline at the tip of South America. During the ice ages, the Patagonian Ice Sheet expanded down from the Andes, covering a full third of Chile at its greatest extent, and glacially gouged out these valleys. They were subsequently drowned by the rising sea levels, becoming an incredibly intricate jumble of rugged little islands, headlands and interlinked channels – almost as if the coastline itself had been shattered by the frost.

When the Portuguese explorer Ferdinand Magellan found a route around the tip of South America in 1520 on the first circum-navigation of the world, he did so along a passage created by these flooded, glacially carved valleys. The narrowest points at the Atlantic entrance to the Magellan Strait were formed by 'terminal moraines' – debris pushed ahead by the glaciers acting like bull-dozers and then, as they began to retreat again with the end of the Ice Age, dumped at their base.[56] The 600-kilometre-long

Magellan Strait was a crucial sailing link between the Earth's two greatest oceans for almost four centuries before the construction of the Panama Canal in 1914. Although it is narrow and difficult to navigate, with unpredictable currents, it is still shorter and (as an inland channel) far more sheltered from storms than the tempestuous passage between the southernmost cape and Antarctica found by Sir Francis Drake in 1578.

Glaciation also had profound implications for the reshaping of North America's geography and the subsequent history of the United States. Here the extensive ice sheet diverted the course of the mighty Missouri and Ohio rivers, and when the glaciation thawed these rivers continued to flow along what had been the edge of the ice sheet. Today, they meet the Mississippi in a huge Ψ shape and offer easy east-west transport right across the interior of the continent. The Missouri in particular reaches over 2,000 kilometres west to the Rocky Mountains. It was this river, previously diverted by the Ice Age, that carried the explorers Lewis and Clark most of the way towards the Pacific coast in 1804, and enabled the establishment of an American presence across the huge tract of Louisiana and the Northwest Territories. Other rivers too were diverted by glaciation, such as the Teays and St Lawrence; without these river transportation routes around the Appalachian Mountains, the original Thirteen Colonies might have remained confined to the Atlantic seaboard.

The Great Lakes of North America too are features left behind by the Ice Age, their deep basins gouged out by the advancing Laurentide ice sheet and then filled with its meltwater as it retreated again around 12,000 years ago.[57] Once they were linked by canals, these extensive waterways became hugely important for inland transport from the Atlantic coast before the construction of long-distance railroads, and saw New York, Buffalo, Cleveland, Detroit and Chicago develop into major commercial centres.[58]

Moraines, rubble ridges 40–50 metres high, can be seen stretching right across the north of the US. Long Island in New York is formed of two long moraine ridges dumped at the head of the Laurentide ice sheet, as is Cape Cod, further up the coast in Massachusetts.[59] Moreover, Boston, Chicago and New York are built on thick deposits left behind by the melting of this ice sheet.[60] We mine these moraines and glacial stream deposits around the world for sand and gravel for the aggregate in concrete, road-surfacing, and base material for foundations or railway tracks. In addition, the frigid frontier of the North American ice sheets drove fierce winds, which picked up fine particles of silt, sand and clay that had been ground out of the bedrock, and deposited them further south to create the fabulously fertile loess farmland of the Midwest.[61]

It's on the other side of the pond, however, that we find perhaps the clearest example of the influence of the ice ages on history.

ISLAND NATION

Half a million years ago, Britain was not an island. It was still part of continental Europe, physically connected to France – like conjoined twins – by an isthmus running between Dover and Calais. This land bridge was a continuation of the hump-shaped geological structure known as the Weald–Artois anticline that stretches from south-east England to north-east France, formed of layers of rock buckled upwards in the same tectonic upheaval that created the Alps when Africa slammed into Eurasia.

The land bridge between England and France was eroded away to sever this connection, and this seems to have occurred in a sudden, catastrophic event. Sonar maps of the English Channel distinctly reveal an unusually straight and wide valley on the seafloor,[62] containing streamlined islands and long, kilometre-wide eroded grooves – clear signs of a huge flood of water coursing over the ground.

As we have seen, during our current era of pulsating ice ages, glaciations have caused the global sea levels to drop over 100 metres. This allowed the shallow continental shelf around the North Sea and the Channel basin to emerge as dry land. During the ice age around 425,000 years ago (five ice ages before the most recent glaciation) a vast lake of water became trapped between the Scottish and Scandinavian ice sheets and the 30-kilometre-wide ridge of rock then still linking England and France. This lake was filled with meltwater from the ice sheets as well as the discharge from rivers like the Thames and Rhine. And with no outlet to escape through, the water rose and rose, until inevitably it began to spill over the top of the land bridge. These colossal waterfalls scooped out vast plunge pools on the channel floor and gouged backwards through the barrier until this natural dam collapsed. The entire trapped lake emptied itself as a catastrophic megaflood, widening the gaping breach in the barrier and carving the landforms on the floor of the Channel we can see with sonar today. This first megaflood 425,000 years ago is thought to have been followed by a second event around 200,000 years ago, and between them they wore away what is now the Strait of Dover, leaving the white cliffs as the stump of the former isthmus. With the subsequent thawing after each ice age and the rise in sea levels during interglacial periods, this passage formed the English Channel (or La Manche, as the French would have it).[63]

Britain had become permanently cut off from Europe.

The formation of the English Channel has had profound ramifications through history for Britain, as well as for Europe as a whole. The Channel has served as a natural defensive moat, protecting Britain throughout European history. The last full-scale invasion, the Norman Conquest of 1066, occurred almost a thousand years ago. Britain was close enough to trade and remain intimately involved in the politics of the Continent, but shielded at the same time.

Throughout the constant squabbling, conflicts and shifting borders of continental Europe, Britain has largely escaped the ravages of war on its own home soils and been able to remain distant and insulated, only choosing to intervene when it was in its interests.[64] In the seventeenth century, for instance, it was spared the devastation of the Thirty Years War, which began as a conflict between Catholic and Protestant European states and ravaged much of Central Europe, causing huge population losses – over 50 per cent in some regions – from the resultant famines and disease. Safe behind its natural moat, Britain's situation contrasts in many ways with that of Germany, bounded to the north by sea and to the south by the Alps, but open on both sides through the European Plain. It is this vulnerability from the lack of natural defences that explains much of the insecurity and military ambitions of the states in this region – the Holy Roman Empire, Prussia, and then Germany as a unified nation.

With clearly defined natural boundaries, and a relatively small extent, England achieved the early unification of feudal fiefdoms into a national identity.[65] It has also been argued that it was this reduced threat of invasion and sense of security from external threats that allowed the progressive dispersion of power away from the autocratic monarch to a more balanced democratic system, beginning with Magna Carta in 1215 and leading to the parliamentary system in place today.[66]

What's more, with no land border to defend, Britain's military expenditure needed to be only a fraction of that of its continental rivals.[67] Britain was instead able to focus its energies on building up and maintaining the Royal Navy, not just for defending the homeland – the defeat of the combined French and Spanish fleets at the Battle of Trafalgar in 1805 that sank Napoleon's hopes of invading Britain is the most striking example – but also to guard its overseas colonies and protect its commercial interests and trade routes, as it developed a seaborne empire that came to supersede those of the Spanish, French and Dutch.

Of course, it's impossible to say with any certainty how European history might have played out had Britain not been an island. What might have happened if the Scottish and Scandinavian ice sheets had never merged to trap the glacial lake that disgorged through the Channel, eroding away the isthmus and opening the Strait of Dover? What if the ice ages had been a little less icy? This is not the place for speculating on counterfactual histories, but thinking about the potentially profound alternative outcomes underscores the importance of geology in how we find the world today. If Britain were still linked to the continent by a land bridge, would the blitzkrieg sweep of the Wehrmacht across Europe also have defeated this last bastion of resistance against Nazi Germany? Would Britain have fallen to Napoleon's Grande Armée in 1805, or would Spanish forces have invaded in 1588 (without the need for an armada)?

It could be argued that the strong island nation has helped maintain a power balance in the Continent's history by resisting invasion and preventing any one power from consolidating a European empire. On the other hand, its geographical isolation has created an island mentality that made Britain often stand aloof and reluctant to enter into closer relationships with its Continental neighbours, despite common interests and a shared fate.

Thus the most recent period of our planet's history has allowed our species to spread around the entire globe, and the lasting impressions the pulsing ice ages left on the landscape have had profound implications for the course of human history. The entire story of civilisation has played out during the current interglacial period, and we will now turn our attention to the planetary forces lying behind this fundamental transition in the human story: the domestication of wild plant and animal species and the emergence of agriculture.

Chapter 3

Our Biological Bounty

Between 20,000 and 15,000 years ago the overlapping rhythms of the Milankovitch cycles started warming the Northern Hemisphere once again. The great ice sheets began to thaw and recede, and the deep freeze of the last ice age drew to a close.[1] In North America, much of the run-off water from the melting ice sheets became trapped behind a ridge of debris deposited at the base of the retreating glaciers. This formed immense meltwater lakes, the largest of which has been named Lake Agassiz after the Swiss-American geologist who first proposed the (at the time) radical notion of a past ice age smothering the Northern Hemisphere. By 11,000 BC Lake Agassiz had expanded to cover almost half a million square kilometres of Canada and the northern United States – an area about the size of the Black Sea. Then, the inevitable happened. The natural dam burst and the huge volume of glacial water disgorged in an immense, surging flood. It ran through the Northwest Territories along the current course of the Mackenzie River and into the Arctic Ocean.[2] This sudden release of the trapped water caused an immediate jump in global sea levels. But it was the effects it had on a culture developing some 10,000 kilometres away in the eastern Mediterranean region of the Levant that were far more profound.[*]

[*] In fact, this episode around 11,000 BC is only one of several occasions when Lake Agassiz drained, as meltwaters accumulated again before breaking through the natural dam once more, each time leading to a sudden leap in global sea levels.[3]

PARADISE FOUND AND LOST

Whilst the ice sheets were retreating, forests expanded again to replace the wide bands of arid steppes and scrubland, rivers swelled and deserts shrank. With the warmer, wetter conditions, lush vegetation proliferated and populations of grazing mammals increased.[4] Springtime was returning to the planet, and our hunter-gatherer ancestors found the going much easier. In the Levant, the land flushed with wild wheat, rye and barley, and recovering woodland. Here a people known as the Natufians emerged, who appear to have formed the first sedentary society in the world, even before the development of agriculture. They settled in villages of stone and wood, gathering the wild cereals along with fruit and nuts from the woodland, and hunting gazelle.[5] If ever there were a hunter-gatherer Garden of Eden, it would have been here.

This golden era didn't last long, however. About 13,000 years ago a sharp climatic jolt, lasting over 1,000 years, struck this region of the Near East and the Northern Hemisphere as a whole. This is known as the Younger Dryas event, and it saw a rapid regression of climate, which over the course of just a few decades returned towards a much colder and drier state. And the cause for this abrupt yank back to ice age conditions is thought to have been the disgorging of Lake Agassiz.

The sudden drainage of this vast lake placed a lid of freshwater on the northern Atlantic, which temporarily shut down the pattern of ocean circulation. Today, the world's oceans operate vigorous conveyor belts of cycling waters that transport heat from the equator towards the poles. This is known as the thermohaline circulation, as it is driven by differences in the temperature and salinity of the seawater. Winds blow the warm surface waters from the planet's midriff towards higher latitudes – we'll return to this in Chapter 8 – sustaining the Gulf Stream, for example, that delivers Caribbean warmth and moisture to Northern Europe. Evaporation along the way turns the seawater more salty, and it also cools on its journey north. Both these effects make the water more dense, so that near the poles it sinks to the ocean floor, and returns towards the equator at depth. The sinking of polar water also draws in more water behind it to maintain the current. But the rapid dumping of a huge amount of freshwater into the North Atlantic from the discharge of Lake Agassiz abruptly stalled the salinity pump of this conveyor belt. The shutting down of the ocean circulation system, which was redistributing heat from the equator, shunted much of the Northern Hemisphere back to the conditions experienced during the height of the Ice Age.[6]

For the Natufians, the environmental crisis of plunging temperatures and declining rainfall saw their homelands reverting to arid, treeless steppes of thorny shrubs, and the abundant wild food sources dwindled before their eyes. It seems that at least some of the Natufians responded by abandoning their fledgling sedentary lifestyle and returning to migratory foraging. But some archaeologists believe that this Younger Dryas event spurred others to turn from their hunter-gatherer ways and instead develop agriculture. Rather than roaming further and further to collect enough food to survive, they brought seeds home and planted them in the ground – the first step of domestication. Plump rye seeds found in the archaeological remains of Natufian villages have been interpreted as signs of this development. The claim is controversial, but if this was the case it would make the Natufians the first farmers in the world. An invention that would for ever change the way we live was born out of the hardship of sudden climate change.[7]

Prompted by a particular chain of planetary events – the disgorging of Lake Agassiz, the stalling of the Atlantic circulation system, and the jolt of the Younger Dryas event – the Natufians may have been the first seed-sowers, but they were already a settled culture, and so perhaps uniquely set up to try this earliest experiment with farming. Yet within a few millennia, as the planet warmed after the last ice age, people around the world came to follow. Between about 11,000 and 5,000 years ago agriculture developed in at least seven different places across the Earth.

THE NEOLITHIC REVOLUTION

While anatomically modern humans had appeared in Africa by around 200,000 years ago, our ancestors only became behaviourally modern between 100,000 and 50,000 years ago. They now possessed the same linguistic and cognitive faculties that we have today, lived in social groups, and crafted and used tools and fire proficiently. They carefully buried their dead, made clothes, and produced expressive artwork in which they portrayed themselves and the natural world around them in cave paintings and bone and stone sculptures. They were adept hunters, they fished, and they gathered a wide variety of edible plants. They had even begun to grind wild grain into flour on simple millstones.[8]

As we saw in the last chapter, from about 60,000 years ago humanity migrated out of Africa and dispersed around the entire globe. But it wasn't until around 11,000 years ago that the first enduring steps towards agriculture and settlement were taken, a transition known as the Neolithic Revolution. The North American ice sheet, though quickly shrinking, still covered more than half of Canada when the first crops were being domesticated in the Fertile Crescent in the eastern Mediterranean, and then shortly afterwards in the Yellow River valley in northern China.[9] Within just a few thousand years our ancestors in several other regions of the world were doing the same. Agriculture also emerged in the Sahel band of North Africa, the lowlands of Mesoamerica, the Andes–Amazon region of South America, the woodlands of eastern North America, and New Guinea.[10] After living 100,000 years through the last ice age as hunter-gatherers, with the warming of the world different peoples around the planet started down the road of agriculture and civilisation that transformed our species for ever.

It's almost as if a starting gun had gone off. What were the planetary forces behind this defining step of human existence?

We can't be certain why people in different places around the world first turned their hand to deliberately sowing seeds and carefully tending plants, beginning the process of domestication and selective breeding of crops. The development of agriculture may have been spurred by a spell of favourable climate that made attempts at farming less risky and more inviting, or conversely by a sudden regional shock of deteriorating conditions – like the the Younger Dryas event – that prompted a settled community to find different ways of feeding itself.[11] But either way the end of the last ice age was clearly instrumental.

The fact that humanity didn't settle down to begin cultivating the land during the Ice Age is perhaps not surprising, although the reason isn't so much the cold conditions. While the northern ice sheets extended far down from the Arctic to smother much of the high latitudes of America, Europe and Asia, conditions were not impossibly cold elsewhere. The temperature around the tropics was only a degree or two cooler than today. And although the Ice Age Earth, as we've seen, was on the whole drier, it was not so arid everywhere as to prevent the development of agriculture.[12] The limiting factor was probably not that the climate was inimically cold or dry, but that it was extremely variable. Regional climate and rainfall could shift suddenly and dramatically.[13] Any Ice Age tribes that attempted precocious experiments with cultivation are likely to have had their efforts snuffed out by just such a rapid fluctuation. Even later in our history, well-established civilisations have collapsed when their regional climate dried out and crippled their agricultural support, such as the Harappans in India, the Old Kingdom in Egypt, and the classical Maya.[14]*

Interglacial periods, such as the one we're living in right now, on the other hand, are marked by their comparatively steady climatic conditions. Indeed, the last 11,000 years of the current Holocene interglacial have been the longest stable warm period

---

* Indeed, it's possible that there may have been earlier instances of settling and farming that faltered without leaving any archaeological traces – false starts in the emergence of civilisation. In particular, any settlements on the Ice Age coastal plains would now be lost beneath the waves after the oceans rose again.[15]

of the past half-million years.[16] And the rise in atmospheric carbon dioxide after the last ice age, which would have invigorated plant growth, was a global effect and so may explain why cultures around the world developed agriculture almost simultaneously.[17] Such stable, warm and wet conditions in regions that reliably produced large-grained grasses would have motivated people to tend a few select species themselves and settle down rather than roam more widely. It seems as though interglacial periods are a prerequisite for farmers.

Let's look in detail now at how we domesticated wild plants and animals, and what determined which species became adopted by humanity.

SEEDS OF CHANGE

The Holocene is the first interglacial period that modern humans have experienced, and almost immediately after it began peoples around the world started developing agriculture. Wheat and barley were first domesticated around 11,000 years ago in the rain-watered, hilly landscape of southern Turkey[18] and then were spread to the plains between the Tigris and Euphrates, a region called Mesopotamia – 'the land between the rivers'.[19] Irrigation was first developed in the Turkish highlands a couple of thousand years later, and then adopted in Mesopotamia 7,300–5,700 years ago to control and distribute the floodwaters of the two rivers.[20] The region curving between Mesopotamia, the Levant and the River Nile is known as the Fertile Crescent: an arc of cultivable land within the otherwise arid environment of North Africa and the Middle East.

In China, millet was cultivated from around 9,500 years ago in the cooler, and seasonally drier, valleys of the Yellow River in the north-west. This millet, and then the soya bean that was domesticated around 8,000 years ago, were grown in the soft, fertile loess soils of the region.[21] Around the same time, rice cultivation began along the Yangtze river in the warmer and wetter tropical region of southern China.[22] Here huge amounts of rice came to be grown in paddy fields and carefully constructed terraces on hillsides, which demanded skilful water engineering to create ponds only a few inches deep in each paddy field that could be drained before harvest.[23]

The crops domesticated in the Fertile Crescent spread to the Indus Valley around 9,000–8,000 years ago, and rice cultivation started in the Ganges Delta, possibly domesticated independently of that in China.[24] In the Sahel, the band of semi-arid climate between the Sahara Desert and the savannah further to the south, the cultivation of sorghum and African rice began around 5,000 years ago, before the continued drying out of the region forced farming communities to migrate to the more humid regions of West Africa.[25]

In the Americas, squash plants were domesticated in Mesoamerica around 10,000 years ago, and maize (corn) was grown in southern Mexico from 9000 years ago; later, beans and tomatoes also became staple crops here.[26] The potato was cultivated in a large number of varieties in the Andes from about 7,000 years ago.[27] In the highlands of tropical New Guinea, the starchy tubers of yam and taro were cultivated between 7,000 and 4,000 years ago.[28]*

* It is intriguing to realise that, had they not unwittingly been saved by people, several of the plants we domesticated would have fallen extinct. The fruit of the wild ancestors of squash, gourds, pumpkins and courgettes, for example, are all repulsively bitter and encased in a hard rind. They naturally relied on large animals like the mammoth and mastodon to break them open and disperse the seeds within. And so when such large beasts died out, these plants were themselves coasting on borrowed time. But about 10,000 years ago, these species were pulled back from the brink when they formed a symbiotic partnership with a new animal species – humans. We domesticated these plants, provided them with new, artificially maintained habitats in our farms and plantations, and changed them over generations of selective breeding to become larger, softer-skinned and more palatable. Avocado and cocoa are also believed to have originally relied on recently extinct large mammals to spread their seeds, and so were saved by humans who adopted these ghost species and became substitute seed dispersers.[29]

By about 5,000 BC, therefore,[30] humanity had learned to domesticate a wide diversity of edible plant species in a variety of climatic zones and landscapes, from the river floodplains of Mesopotamia to the high terrain of the Peruvian Andes and the tropics of Africa and New Guinea. By far the most significant plants we cultivated are the cereal crops. Grains like wheat, rice and maize, along with millet, barley, sorghum, oats and rye, have supported millennia of human civilisation. And the three most important systems of agriculture that have spread across most areas of the globe are wheat originating in the Fertile Crescent, rice in China, and maize in Mesoamerica.[31] Today, these three cereals alone provide around half of all the human energy intake around the world.

Cereal crops are all species of grass. The astonishing truth is that we are no different from the cattle, sheep or goats that we leave out to pasture – humanity survives by eating grass.

Many grasses are hardy species of plant that can colonise land after pre-existing forest has died back with increasingly arid conditions, after fire has torn through an area, or in fact after any other disruption to the established ecosystem. Their survival strategy is to grow fast and put most of the energy they gather from the sun into their seeds, rather than building stout frames like trees – which is what lends them to cultivation. And this is the fundamental ecological reason that so many of us eat a slice of toast or a bowl of cereal for breakfast – wheat bread, cornflakes, rice crispies and porridge oats are all derived from fast-growing species of grass (and cereal crops of course also form the staple of other meals too).

But to make use of grassy cereal crops we are still faced with a biological problem. We don't have four stomachs like a cow that would enable us to break down tough plant matter to release its nutrients. We therefore picked plant species that produce a concentrated dollop of energy in their grain – which,

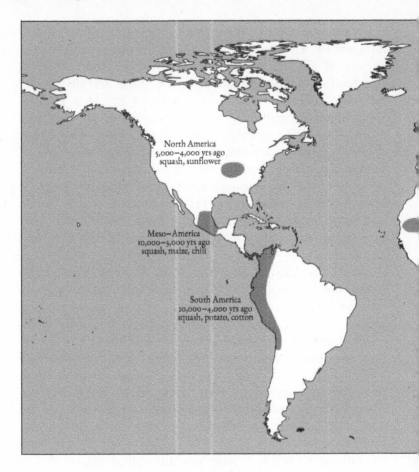

North America
5,000–4,000 yrs ago
squash, sunflower

Meso–America
10,000–3,000 yrs ago
squash, maize, chili

South America
10,000–4,000 yrs ago
squash, potato, cotton

Origins of crop domestication.

botanically speaking, is a fruit – and applied our brains to the problem rather than our stomachs. The millstone we use for grinding grain into flour (and the mechanisms we invented through history to drive its rotation, like the waterwheel or the windmill) is a technological extension of our molar teeth.

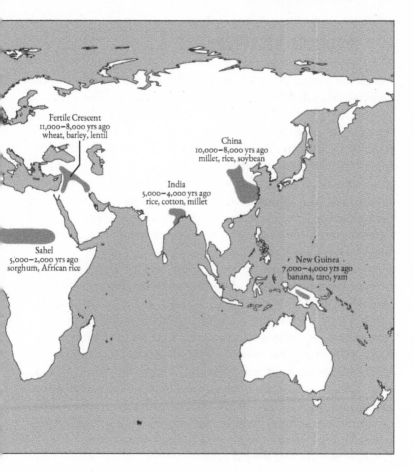

And the oven we use for baking that flour into bread, or the pot we use for boiling rice and vegetables, are like external pre-digestive systems. We have applied the chemically transformative power of heat and fire to break down the complex plant compounds so that we can absorb the nutrients.

POINT OF NO RETURN

The development of agriculture offered huge advantages to the societies that adopted it, despite the continuous labour involved in working the land and nurturing the crops. Settled peoples are capable of much faster population growth than hunter-gatherers. Children don't have to be carried long distances and babies can be weaned off breast milk (and fed with milled grain) much earlier, which means that women can give birth more often. And in agricultural societies, more children are an advantage for they can help care for crops and livestock, mind their younger siblings and process food at home. Farmers beget more farmers very effectively.[32]

Even with primitive techniques, an area of fertile land can produce ten times more food for humans when under cultivation than when used for foraging or hunting.[33] But agriculture is also a trap. Once a society has adopted cultivation and its numbers have grown, it's impossible to revert to a simpler lifestyle: the larger population becomes entirely dependent on farming to produce enough food for everyone. There's no turning back. And there are other consequences too. High-density, settled populations supported by farming soon develop highly stratified social structures, resulting in reduced equality and a greater disparity in wealth and freedom compared to hunter-gatherers.[34]

When farmers first moved down from the hills of modern-day Turkey into the plains of Mesopotamia in the sixth millennium BC, bringing with them their domesticated cereal crops, the Earth was entering the warmest, wettest phase of the Milankovitch cycles. The marshy ground of lower Mesopotamia was extremely fertile, its thick alluvial soil eroded out of the highlands to the north and deposited by the rivers as they flowed into the Persian Gulf. (Mesopotamia runs along a tectonic trough, as we saw in Chapter 1.) Productive agriculture fuelled a population boom, but by 3800 BC the climate was becoming cooler again and the rains less reliable: the fecund land between the rivers began to

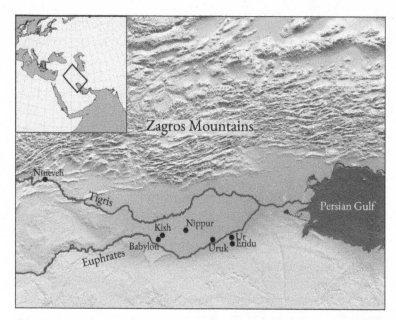

Mesopotamia lies in the tectonic trough formed alongside the Zagros mountains.

dry out. In response village farmers pooled their resources and manpower and congregated into larger and larger settlements from where they could operate more extensive irrigation systems.[35] Building and maintaining these canals for both agriculture and transportation required, and in turn fostered, a centralised administration and increasingly complex systems of social organisation.[36] So it was here in Mesopotamia that agriculture gave birth to the world's first urbanised society. By 3000 BC more than a dozen large cities had been founded,[37] their names still persisting in our cultural memory: Eridu, Uruk, Ur, Nippur, Kish, Nineveh, and later Babylon. The land between the rivers had become a land of cities, known to its inhabitants as Sumer.[38] By 2000 BC, 90 per cent of the Sumerian population were living in cities.*

* The Sumerian cities, fed by the fertile alluvial soil, were also largely built of the river mud beneath their feet, as we'll see in Chapter 5.

The emergence of civilisation in ancient Egypt is also believed to have been a product of climate change. During previous interglacials North Africa has been substantially wetter, dotted with large lakes and sporting extensive river systems, making the Sahara green with grasslands and woods.[39] Roving tribes hunted across this landscape of savannah and woodland, and fished the lakes and rivers. The only signs remaining today of this thriving wildlife in the region are the rock paintings left by the hunters, depicting crocodiles, elephants, gazelles and ostriches.[40]

This climate optimum was not to last, though. As Mesopotamia began to dry, the monsoons also retreated from North Africa. The remaining pockets of surface water in the Sahara soon disappeared and the area desiccated rapidly by the end of the fourth millennium BC.[41] Humans living here saw the environment around them degrading as it slipped into its current hyper-arid state. At first, they may have survived around the remaining oases, but as the region continued to desiccate they abandoned the dying land and retreated into the Nile Valley. Egypt had inherited the crops and animals domesticated in the Near East, with agricultural villages first emerging in the delta, and then along the Upper Nile, from about 4000 BC. Around 3150 BC, just as the Sahara finally dried out, the region was unified under the rule of the dynastic pharaohs.[42] The process of increasing population density, social stratification and state control that marked the beginnings of Egyptian civilization was thus driven by climate refugees from the desertifying Sahara crowding into the narrow Nile valley.[43]

Ancient Egypt offers perhaps the clearest case of how the development of a civilisation is influenced by the combination of constraints and opportunities presented by its geographical setting and climate. A ribbon-like oasis running through the desert, the Nile's reliable summer floods revitalise the plains either side of its course with mineral-rich sediment eroded out of its headlands in the highlands of Ethiopia. The mighty Nile also provided a simple means of transport. The prevailing north-east trade winds blow reliably in north African latitudes – we'll return to them in Chapter 8 – which means that boats can sail south to Upper Egypt; and the Nile's gentle current then allows an easy return downriver with the flow. This natural two-way transit system[44] not only made possible the ready transport of grain, wood, stone and military forces, but the ease of communication along the whole length of Egypt helped consolidate the unified state.

Egypt is well defended by natural barriers of inhospitable desert on either side of the Nile and so was able to resist invasion for most of its history.[45] But this containment also prevented Egypt from extending its territory into a sprawling empire; apart from a late second millennium BC expansion along the Levant coast, Egypt remained a regional power along the Nile. While the river valley was fabulously productive for grain – it helped feed the city-states of ancient Greece and later became the breadbasket of the Roman Empire – it lacked an abundance of trees. Cedar timber was imported from the Levant, but it was too expensive to build a large navy to project Egyptian power across the Mediterranean or beyond the Red Sea.[46]

It was this combination of environmental advantages, simple internal transport, the ecological sustainability of agriculture offered by the Nile, and the natural defensive barrier of the enveloping desert that created the stable and long-lived Egyptian civilisation.[47] Above all, it was the river that bestowed prosperity on the region. As the Greek historian Herodotus wrote in the fifth century BC, Egypt is 'the gift of the Nile'.

So within a few centuries of the first Sumerian urban centres, cities and systems of greater social organisation were also emerging in the Nile, as well as the Indus and Yellow River valleys.[48] Bountiful agriculture produced grain surpluses to feed these ever more populous settlements, and rulers coordinated the labour of growing workforces to construct impressive civic engineering projects like expansive irrigation systems, roads and canals, to further increase food production and its distribution. And within the cities, the proportion of the population released from the need to produce food could specialise in other skills: carpentry, metal-working, or even investigating the natural world. Stored surpluses of grain also supported large armies and generals soon consolidated the first empires of the world.

TAMING THE WILD

The birth of civilisation did not just depend on the cultivation of plant species. It also relied on turning wild animals into livestock.

The domestication of the first animal pre-dates humanity's settling down. Dogs were tamed from wolves by European hunter-gatherers during the last ice age more than 18,000 years ago,[49] to help with the hunt or warn people of approaching predators. But the majority of the animals on our farms today were adopted into the care of humans much more recently, alongside the earliest cultivation of crops. Sheep and goats were domesticated in the Levant a little over 10,000 years ago – sheep in the foothills of the Taurus Mountains and goats in those of the Zagros mountain range.[50] At around the same time cattle were domesticated from the wild aurochs in the Near East and India. The pig was domesticated in Asia and Europe between 10,000 and 9,000 years ago, and the chicken in South Asia around 8,000 years ago. In the Americas, the llama was domesticated in the Andes about 5,000 years ago, and the turkey in Mexico 3,000 years ago.[51] The poultry species domesticated in the Sahel was the guinea fowl.[52]

In all these cases, domestication would have come at the end of a long process of cohabitation in the wild. People would not have invested time and energy into breeding, feeding, rearing and protecting animals, had they not already been intimately familiar with their habits and uses. So over a long road of humanity's interaction with the animals around us, we turned from scavenging carcasses, to hunting and then to husbandry.

As we have seen, the transformation of wild plant species into cultivated crops allowed much greater food production, even if it came at the cost of a higher investment of time and effort. At the same time, the domestication of these animal species offered a reliable source of meat without the bother of a long hunt. But the domestication of animals also provided other opportunities that had not been available to roving hunter-gatherers. From an animal that has been killed you can extract its meat, blood, bones and hide. These are all extremely useful products for food, tools and shelter, but you only get them once. If, on the other hand, you care for the animals, rearing and protecting them, you can ensure a more reliable supply of these products as you cull your herd. And once live-stock have been domesticated and are tended throughout their life, it is also possible to extract other useful products and services from the animals on a continuous basis, which you simply cannot exploit from wild beasts. Animal husbandry offers you completely new resources. This has been termed the 'secondary products revolution'.[53]

Milk is one such new resource. First goats and sheep, then cattle, and in some cultures even horses and camels, were milked for human consumption – human mouths essentially replacing those of the animal's own young. Milk provides a reliable source of nutrition – it is rich in fat and protein as well as calcium – and products made from it such as yoghurt, butter and cheese preserve these nutrients for long periods of time. Over its entire lifetime, a mare kept continually lactating supplies about four times more energy than what its meat at slaughter would provide.[54] Only human populations native to Europe, Arabia, South Asia and western Africa are able to digest fresh milk, however.[55] They have evolved so that the milk-digesting enzyme in their gut, which in other mammals only exists in babies, continues to be produced throughout their entire adult life. This is one of the clearest examples of humanity co-evolving with the animal species that we domesticated and selectively bred for our own purposes.

Wool, too, can be continually harvested from domesticated livestock. Wild sheep are hairy, with only a thin undercoat of short, fluffy fibres. Over generations of selective breeding humanity has emphasised this undercoat to provide the wool that was first plucked and then shorn for weaving into clothing, a development that occurred between 5,000 and 6,000 years ago.[56] Llamas and alpacas served an equivalent function in South America.

And the domestication of large animals provided another important resource unavailable to hunter-gatherer societies: their muscle power as beasts of burden for transport and traction. The first species used to carry loads as pack animals was the donkey, but it was superseded by the horse, the mule (the infertile hybrid of horse and donkey) and the camel, all of which can carry larger loads further. Cattle were the first animals used for providing traction – pulling ploughs or wagons – as it is relatively easy to hitch a yoke to their horns; oxen (castrated bulls) in particular are strong but placid.[57] The application of animal traction enabled the transition from agriculture powered by human muscle, with farmers using small hand tools like the hoe or the digging stick, to the use of ploughs. Livestock pulling ploughs gave another boost to food production. It also allowed marginal land, which had previously been considered too poor quality, to be turned to farming. Pack animals carrying goods over uneven ground or traction animals hauling carts and wagons over flat plains greatly increased both the bulk and the range of goods that could be transported, and so were hugely important for establishing long-distance overland trade routes. In addition, horse-drawn chariots revolutionised warfare in Eurasia in the second millennium BC; and later, once larger and stronger horses had been selectively bred and horseback riding became possible, cavalry became the most effective weapon of war.

Domesticated animals are especially beneficial when used in combination with each other. This was particularly important for nomadic, pastoralist societies: in regions with little arable land, people adopted a lifestyle supported almost entirely by large herds of livestock, with which they roamed between pastures. Animals such as sheep, goats and cattle are like food-processing machines. They thrive on the plains of grass, unsuitable for human consumption, and transform it into nutritious meat, marrow and milk. They also produce wool, felt and leather for clothing, bedding and tents. For pastoral societies these animals provide the very foundations of survival and a source of wealth that can be traded.[58] Mounted herders riding swift horses are able to control huge flocks grazing over enormous areas of land, greatly amplifying the animal resources that pastoralists can maintain. And the bulk transport provided by oxen-hauled wagons, serving as mobile homes, enabled family groups to roam far and wide with their herds. It was this integration of herded livestock, horseback riding and animal traction that opened up the vast grasslands of central Eurasia as a habitat for pastoral nomads. The interaction – and often violent conflict – between these nomadic tribes living across the breadth of the steppes and the settled agrarian societies around its margins played a pivotal role in the course of Eurasian history, as we'll see in Chapter 7.

The use of animal muscle power greatly expanded the capabilities of human societies – long-distance trade and travel through different environments became possible with horse, mule and camel, and strong but slow animals like oxen or water buffalo provided traction for pulling wagons and ploughs. And with the innovation of the collar harness in fifth-century China, horses could be used for traction as well – a development that greatly increased medieval agricultural productivity in the heavy soils of northern Europe. Domestication of these animals to replace human muscles was the first stage in the progressive story of humanity's marshalling greater and greater energy sources.[59] Animal power reigned supreme for driving civilisation for around six millennia, before the introduction of fossil energy during the Industrial Revolution, when coal-fired steam engines began to propel trains and ships, and later the internal combustion engine, fed on liquid fuels refined from crude oil, enabled us to cover vast distances at astonishing speed.

Let's now turn to the planetary forces that created these crucial animal and plant species we came to domesticate.

SEXUAL REVOLUTION

Our modern world of glittering skyscrapers and intercontinental flights still feeds itself on the grass species domesticated by our ancestors around 10,000 years ago. These cereal staples provide the majority of our daily energy needs, but of course humankind does not live on bread alone. Our diet also includes many other varieties of fruit and vegetable. Despite this apparent diversity, however, virtually all the plants we consume are members of one particular group, known as the angiosperms. I'll explain their characteristics in a minute, but first let's look at earlier forms of vegetation to put the astonishing evolutionary innovations of the angiosperms into perspective.

The primeval trees of the Carboniferous Period, which would provide the vast reserves of coal fuelling the Industrial Revolution, and still give us a third of the energy we consume today,[60] were a kind of plant known as spore-forming. Like ferns today, they reproduced by releasing spores on the wind which, if they fell on accommodating ground, germinated and grew into a tiny, green, leafy plant in their own right, but with only one-half of the full set of genetic material. It is this separate plant stage that had the equipment for sex, and they produced sperm that swam through films of water in the soil to an egg cell on a nearby plant. Once fertilised to reconstitute a complete double set of chromosomes, the egg then grew into a new full-sized tree. This seems a truly bizarre way of reproducing. It is as if humans procreated by spraying their sperm and eggs onto the ground in front of them, which each developed into a miniature version of themselves, and which then had to mate with each other to create an adult person. Moreover, this reproductive strategy worked fine for the spore-forming plants in the swampy basins of the Carboniferous, but they were biologically restricted to soggy soils by this alternating life cycle.

Gymnosperms – plants with 'naked seeds' – emerged at the end of the Carboniferous and developed into all the evergreen conifers familiar to us today, including fir, pine, cedar, spruce, yew and redwood. They evolved to effectively suppress that intermediate phase of the life cycle. Once pollinated, gymnosperms produce seeds that are exposed on the scales of their cones. The seeds fall to the ground, safe within their protective casing and containing a small provision of stored energy, and wait for the right conditions to sprout. This evolutionary innovation released plants from the wetlands. (In some ways it's analogous to the evolution of reptiles which, unlike amphibians, didn't need to return to the water to reproduce.) As the gymnosperms spread around the world other plant species became either literally over-shadowed – bracken and other ferns mostly survived in the gloomy understorey within forests – or, like ginkgo in central China, continued to thrive only in isolated pockets. Gymnosperms are still very common today, growing as the dense conifer forests of spruce, pine and larch in the taiga ecosystem that stretches between the arctic tundra and the grasslands of the North American prairies and Eurasian steppes. They have been important through human history as sources of softwood for construction timber or pulp for paper, and feature as a minor part of our diet, for example in the form of pine nuts toasted and tossed into a salad or ground into pesto.

The naked-seed gymnosperms ruled the Earth's vegetation for around 160 million years, but it is angiosperms that dominate the plant world today, both in terms of their rich diversity of species and the range of different habitats around the planet they've come to master: deciduous woodlands in temperate regions, tropical rainforests, vast plains of grass across drier regions, and cacti in deserts. Angiosperms have taken their sex lives to an even higher level of refinement. Their eggs are not left naked but are contained in a special organ, originally adapted from a curled-up leaf, within which the seeds then develop – angiosperm means 'encased seeds'.[61]

A far more noticeable defining feature of the angiosperms, however, is the way they adorn and advertise their sex organs with flamboyant displays in the development of the flower. This evolutionary invention enabled angiosperms to recruit a huge range of insects – as well as birds, and some bats and other mammals[62] – to help them transfer pollen from one plant to another. The first flowers were probably simply white, but as these plants and their pollinators developed together – one of the greatest stories of co-evolution in the history of life on Earth – the world exploded in a profusion of floral colours and heady scents. The specialised sex organs of flowering angiosperms not only allowed them to co-opt animals into helping them with their reproduction, but the ovary containing the seeds also developed into fleshy means to help them disperse: it produced fruit.

By the late Cretaceous, the last period of the dinosaurs, the plant world on our planet would have already begun to look pretty similar to today, with sycamore, plane, oak, birch and alder tree families well established. But there was one glaring exception. The open, unforested plains in the drier areas of the continents would still have seemed eerily different. Although early forms of heather and nettles existed, grass species didn't evolve until the end of this period.[63] The dinosaurs roamed over terrain entirely devoid of grass.

Our evolution as primates and our development as hunter-gatherers depended on the fruit, tubers and leaves of angiosperm plants. And the agriculture we adopted is also almost entirely reliant on angiosperms. Cereals are angiosperms: in fact, the grain we harvest is botanically the fruit of the grass plant.[64]

Signs of grass first appear in the fossil record from about 55 million years ago, but with the persistent cooling and drying of the planet through the Cenozoic era, grass-dominated ecosystems became established in many parts of the world between 20 and 10 million years ago.[65] So not only was our own evolution driven by the aridification of East Africa, but the cooling and drying of the world as a whole created the conditions for the spread of the plants we would come to domesticate as staple crops for feeding our civilisations through history. And virtually every other plant we eat is also a member of one of eight different families of angiosperm.

After the grasses, the second most important family are the legumes, which include peas and beans, soya beans and chickpeas, as well as the alfalfa and the clover we feed to our livestock. The brassicas include rapeseed and turnip, and a single species of this family, a weedy mustard plant, was transformed by accentuating different features of the plant through selective breeding to give us cabbage, kale, Brussels sprouts, cauliflower, broccoli, and kohlrabi.[66] Other angiosperm groups include the nightshade family of potatoes, peppers and tomatoes; the family of gourds, pumpkins and melons; and the parsley family that also includes parsnip, carrot and celery.

Most of the fruit we consume comes from either the rose family (such as apples, pears, peaches, plums, cherries and strawberries) or the citrus family (oranges, lemons, grapefruit, kumquat). The family of palm trees have also played an important role in history, giving us the coconut and, more influentially, the date, which served as a light and concentrated food source for the trade caravans crossing the deserts of the Middle East.

Across these families of angiosperms we eat different parts of the plant. We cherish fruit which were evolutionarily designed by angiosperms to be attractive and tasty to animals to help them spread their seeds. Plants also create internal energy stores to power their growth the following spring and these are the root and stem vegetables we cultivate. Swollen roots include cassava, turnips, carrots, swedes, beets and radishes, and the tuber of a potato or yam is the swollen section of the plant stem. We eat the leaves of cabbage, spinach, chard and pak choi, as well as other salad plants and herbs; and the cauliflower and broccoli we consume are in fact immature flower heads. So overall, not only do we feed ourselves on grass, but also relatives of the rose bush and deadly nightshade. And beyond providing food angiosperms also give us fibres, such as cotton, flax, sisal and hemp, and a range of natural medicines.

THE CIVILISATION APP

While we cultivate and munch our way through a pretty broad range of different kinds of angiosperm plant, we have been far more limited in the sorts of large animals that we domesticated: we've selected them exclusively from just two categories of mammal.

The first true mammals emerged around 150 million years ago, but it was the mass extinction of species 66 million years ago, wiping out the dinosaurs, that allowed our mammalian ancestors to spread into the niches now left vacant by reptiles. The three major orders of mammals that dominate the world today, however, did not emerge and begin diversifying until 10 million years later. These are the artiodactyls, perissodactyls and primates – collectively known as APP mammals.[67]*

We ourselves belong to the primates, as we saw in Chapter 1, and so they require no further introduction. Artiodactyls and perissodactyls, on the other hand, may sound like alien species, but you are intimately familiar with them. In fact, you could argue that they provided the very basis for human civilisation. They are the two branches of ungulates, or hoofed

---

* In the hierarchical system that we developed for classifying different organisms, artiodactyl, perissodactyl and primate are known as different orders. They all fall within the class of mammals (and ultimately the kingdom of the animals), and within them they hold individual species – for example, the cow (*Bos taurus*).

mammals. The artiodactyls are the even-toed or cloven-hoofed ungulates; the perissodactyls are the odd-toed ungulates.

The even-toed artiodactyls include pigs and camels, as well as all the ruminants: antelope, deer, giraffe, cow, goat and sheep. Ruminants deal with the challenge of breaking down tough grass by regurgitating the cud to chew it again, and then use bacteria in the first of four compartments in their stomachs, the rumen, to ferment the plant material and help break it down chemically, before it is passed through the rest of the digestive system to absorb the nutrients. (As we saw earlier, humanity found technological solutions to the same biological problems.) Artiodactyls are the dominant large herbivorous animals in the world today. Their cloven hoof is made up of two toes, which correspond to the third and fourth fingers on your hand.*

The odd-toed perissodactyls include horses, donkeys and zebras, as well as tapirs and rhinos. Perissodactyls have either three toes like the rhinoceros or just one like the horse. In effect, horses gallop around on the same finger you would use to flip someone the bird. In contrast to ruminants, they are hind-gut fermenters with a simpler stomach. They host bacteria to ferment and help release the nutrients from the vegetation in a greatly enlarged pouch in their intestine, called the cecum.†

---

* Artiodactyls haven't always been herbivorous. Twenty-five million years ago Archaeotherium, related to the hippo and the whale, roamed North America, and this fanged, cow-sized predator may even have attacked rhinos.

† The distinction between the artiodactyls and perissodactyls isn't just an arcane detail of evolutionary biology but has become deeply embedded within religions. The Torah permits Jews only to eat mammals that both have cloven hooves and also chew the cud. Thus, evolutionarily speaking, only members of the ruminant branch of the artiodactyl ungulates are considered kosher or fit for consumption. Jewish scripture (Deuteronomy 14:6–8) also deals specifically with the camel, which despite being anatomically even-toed and also chewing the cud is proscribed as unclean (its feet have padded soles of hardened skin that hide the hoof).[68] The Islamic faith, on the other hand, is less restrictive on the consumption of different mammalian species. The Quran only specifically excludes the meat of pigs, and in contrast to Judaism, camel is generally considered to be halal.[69]

It is astonishing that the vast majority of the large animals that we domesticated over the last 10,000 years, and which human civilisation came to depend upon for their meat, secondary products and muscle power, are all members of just one group of mammals. But there's something equally fascinating and profound about how these ungulates first emerged.

A FEVER OF THE WORLD

The surprising fact is that the artiodactyl and perissodactyl orders, along with the primates, all emerged suddenly within a period of about 10,000 years, in a burst of evolutionary diversification that occurred 55.5 million years ago. It turns out that both our ancestors who would eventually evolve into *Homo sapiens* in East Africa and the groups of animals which became so vital for domestication and the development of civilisations, all appeared in the same blink of planetary time. And the event that seems to have triggered the rapid emergence of these crucial APP mammals was a singular planetary spasm – an extreme spike in the world's temperature.[70]*

This exceedingly rapid heating of the world's climate marks the boundary between the geological epochs of the Palaeocene and the Eocene, and so is known as the Palaeocene–Eocene Thermal Maximum – PETM for short. Over a very brief geo-logical span of less than 10,000 years, massive amounts of carbon (carbon dioxide, CO_2, or methane, CH_4) were injected into the atmosphere, creating a powerful greenhouse effect, and the global temperatures jumped rapidly by 5–8 °C in

---

* It's important to be clear that the first members of these new orders of mammals emerged at this time 55.5 million years ago, but the species within these orders that we are familiar with today did not evolve until much more recently – about two million years ago for the wild ancestor of the cow, for example.

response.[71] This temperature spike made the world the hottest it has been for the past few hundred million years.[72]

Despite this huge jolt to the environment, no mass extinction on the scale of the end-Cretaceous or end-Permian (see pages 40, 141–3) was triggered, although the ecosystems of the world were utterly transformed. Tropical conditions extended all the way to the poles, with broad-leafed trees, crocodiles and frogs all thriving within the Arctic.[73] The PETM caused the disappearance of some deep-sea amoeba, called foraminifera, that were unable to cope with the warmer waters and reduced oxygen at depths,[74] whereas plankton such as dinoflagellates bloomed in the balmy sunlit surface of the oceans. The global environmental disruption of the PETM also drove rapid evolution in many animals,[75] and in particular this temperature spike seems to have ushered the emergence of the new APP orders of mammals.[76]

We would expect that the rapid heating of Earth's atmosphere was the result of volcanic activity, as happened on numerous occasions in our planet's history. But the curious thing is that the cause of most of this enormous and sudden release of carbon that triggered the temperature spike wasn't volcanic – it was biological.*

It is thought that an initial volcanic eruption released enough carbon dioxide to warm the oceans sufficiently to destabilise underwater deposits of a kind of ice called methane clathrate. Clathrate ice forms under the cold and high-pressure conditions

---

* We know about this from measuring the carbon contained in rocks on the sea floor. Carbon atoms are present as several variants with different atomic weights, known as isotopes. Light carbon is preferentially captured by key biochemical reactions, and so the molecules in living organisms, or the carbon dioxide or methane they release, contains more of the light carbon. When scientists analysed the carbon isotopes within limestone rocks laid down on the sea floor during the PETM (which is a way of measuring the atmosphere at that time) they found a huge jump in the proportion of light carbon. This means that the carbon dioxide or methane gas that had surged into the atmosphere to cause this temperature spike must have originally come from life.

of the sea floor and traps methane gas originally produced by decomposition bacteria. But if these clathrates are warmed they break down and release the trapped methane to bubble up through the water and into the atmosphere. Methane is one of the most powerful greenhouse gasses – its heat-trapping effect is over 80 times stronger than carbon dioxide – so the first methane that was unleashed caused further warming which destabilised even more clathrate ice in a diabolical feedback process. Alongside the clathrate ice, more greenhouse gas probably belched out when the permafrost in Antarctica began to thaw and wildfires became more frequent in the warming climate.[77] The initial volcanic eruption was like the detonator that set off the main explosive charge of biological carbon release, resulting in the sweltering climate of the PETM.

Although severe, the temperature spike was very brief in geological terms: the atmosphere and the global climate returned to their earlier levels again within 200,000 years or so.[78] Yet this global warming – a short, intense fever of the world triggered by a great methane flatulence of the oceans – led to the emergence of the three orders of mammals most fundamental for all human history. Artiodactyls, perissodactyls and our own group the primates all appeared suddenly right at the beginning of the PETM, and then rapidly dispersed across Asia, Europe and North America.[79]

If this extreme temperature blip drove the emergence of the APP orders, it was the global cooling and drying over the last few tens of millions of years that created the ecosystems the artiodactyls and perissodactyls came to dominate. As the grass-lands spread around the desiccating continents, the herbivorous ungulates followed and diversified into a large number of different species, including the ancestors of our cows, sheep and horses. So the grasslands that supplied the cereal crops we came to cultivate also provided the evolutionary theatre for the emergence of the large ungulate animal species that we domesticated. But when the world emerged from the last ice age, and human communities around the globe began to settle down and domesticate the wildlife they found around them, both cereal and ungulate species were not evenly distributed around the planet. And this had profound implications for the subsequent course of civilisations.

THE EURASIAN ADVANTAGE

Of the roughly 200,000 plant species in the natural world, only a couple of thousand are suitable for human consumption, and just a few hundred of these offer potential for domestication and cultivation. As we saw earlier, the staples that have supported civilisations across the planet throughout history are cereal crops, but the wild grass species that these cereals were domesticated from were not uniformly spread around the world. Of the fifty-six grasses offering the largest, most nutritious seeds, thirty-two grow wild in south-west Asia and around the Mediterranean, six are found in East Asia, four in sub-Saharan Africa, five in Central America, four in North America, and only two each in South America and Australia.[80]

Thus from the very beginnings of agriculture and civilisation, Eurasia was richly endowed with wild grass species amenable to domestication by humanity and suitable for supporting growing populations. And not only was Eurasia by chance blessed with this biological bounty, but the very orientation of the continent greatly promoted the spread of crops between distant regions. When the supercontinent Pangea fragmented, it was torn apart along rifts that just so happened to leave Eurasia as a broad landmass running in an east–west direction – the entire continent stretches more than a third of the way around the world, but mostly within a relatively narrow range of latitudes. As it is the latitude on the Earth that largely determines the climate regime and length of the growing season, crops domesticated in one part of Eurasia can be transplanted across the continent with only minimal need for adaptation to the new locale. Thus wheat cultivation spread readily from the uplands of Turkey throughout Mesopotamia, Europe and all the way round to India, for example. The twin continents of the Americas, by contrast, though joined by the bridge of the Panama Isthmus, lie in a north–south orientation. Here, the spreading of crops originally domesticated in one region to another entailed a much harder process of re-adapting the plant species to different growing conditions. This fundamental distinction in the layout of the Old World versus the New, itself born from plate tectonics and the aimless wandering of the continents into their current configuration, gave the civilisations of Eurasia a great developmental advantage through history.[81]

The distribution of large animals around the world was equally uneven, and here societies across Eurasia received another advantage. The attributes of a wild animal that make it amenable to domestication by humans include offering nutritious food, a docile nature and lack of inherent fear of humans, a natural herding behaviour, and the ability to be bred in captivity. Yet only a relatively small number of wild animals qualify on all these factors.[82] Of the 148 species of large mammals around the world (heavier than 40 kilogrammes), 72 are found in Eurasia, of which 13 were domesticated. Of the 24 found within the Americas, only the llama (and its close relative the alpaca) was domesticated in South America. North America, sub-Saharan Africa and Australia completely lacked domesticable large animals. The five most important animals through human history – the sheep, goat, pig, cow and horse – as well as the donkey and the camel that provided transport in particular regions, were present only in Eurasia, and within a few thousand years of their domestication had spread across the continent.[83] It is the large mammalian species that have proved most influential throughout history, not only for their meat, but also for their secondary products (milk, hide and wool), and their muscle power.

Equids (species related to horses) evolved in the grassy plains of North America, but by the end of the last ice age the only four groups of equids to survive were all found in Eurasia:

onagers in the Near East, asses in North Africa, zebras in sub-Saharan Africa, and the horse in the Eurasian steppe belt. Similarly, the ancestor of the modern camel – alongside the horse the other animal serving the crucial function of carrying packs or human riders over long distances – lived in the cold climate of the Canadian high arctic and crossed the Bering land bridge into Eurasia with the lower sea levels of a past ice age. The two-humped Bactrian camels in Asia are direct descendants from these American immigrants, and in the hotter deserts of Africa and Arabia the single-humped dromedary evolved to minimise its surface area and thus water loss. These camels became the backbone of the long trade routes through the Sahara, Arabian peninsula and the deserts along the southern margin of the Asian steppe belt. Camelids also migrated across the Isthmus of Panama into South America and developed into the llama and alpaca, but as a beast of burden the llama cannot carry much more than a man, and alpacas were used only for their fleeces.[84]

The great irony of the biological impediment faced by the American civilisations is that these two groups of animals which became so central to transportation and trade across Eurasia had in fact evolved in the Americas and then migrated into Eurasia along the Bering land bridge.[85] But both the horse and camel subsequently died out in their homeland, probably due to over-hunting from early humans who crossed the same land bridge in the opposite direction during the most recent ice age. The first Americans had unwittingly hobbled the future development of civilisations across their continent.

Donkey, horse and camel became critical to the travel and trade routes across the steppes, deserts and mountain passes of Eurasia, Arabia and Africa, greatly empowering the economies and enabling the transfer of people, resources, ideas and technologies across the Old World. The Americas, on the other hand, were biologically impoverished and could not benefit from these revolutions. The camel never returned to the Americas in any significant numbers, but the horse was brought back to its native lands with the Spanish conquistadores in the early sixteenth century. And when contact between the two worlds was renewed in the 1500s, it was European states, the inheritors of this accumulated Eurasian bounty, that came to dominate the cultures of the Americas.[86]

When humanity emerged in the Cenozoic, the age of 'new life', we entered a world characterised by angiosperms and mammals – plants with encased seeds and animals with breasts. But within these broad categories, we have on the whole been surprisingly selective in the species we came to domesticate. Civilisations throughout history have been fed on a staple diet of cereal crops, derived from wild grass species that proliferated around the world as the climate cooled and dried over the past few tens of millions of years. The spread of these grasslands also drove the diversification of the ungulate species that we came to domesticate, providing us with a reliable source of meat, milk and wool, transport and traction power. But when humanity became able to settle as farmers and start down the path of civilisation soon after the end of the last ice age, the uneven distribution of domesticable plant and animal species around the world, as well as the fundamental orientation of the continents, came to exert a deep influence on the patterns of history.

Many of the earliest civilisations to emerge did so along the banks of great watercourses like the Tigris and Euphrates, Indus, Nile and Yellow rivers. They provided the lifeblood for reliable agriculture and the first cities, and political power often arose from the centralised control of their waters for irrigation. Successful agriculture is utterly reliant on intercepting fresh water as it cycles around the world – evaporating from oceans, falling as rain, percolating underground and then flowing back to the sea. Rivers are often the most reliable stage of this water cycle, and they remain critical for feeding many people around the world today. Industrialised agriculture has been refined to the point where it now supports more than 7.6 billion people. Today over 40 per cent of the total global population lives in

India, China and Southeast Asia, and this brings us to the central geopolitical significance of Tibet.

THE WATER TOWER

China has controlled the Tibetan plateau at various periods through its history, such as during the Mongolian Yuan dynasty in the thirteenth century and the Qing dynasty from the early eighteenth century. In recent times, the People's Republic of China under Mao Zedong annexed Tibet in 1951, and after an uprising in 1959 the country's religious leader, the Dalai Lama, fled to India, where a government-in-exile is keeping the independence movement alive in the international eye.

China has two major strategic reasons for wanting control over the Tibetan plateau. The first is military: ensuring that India does not try to secure a commanding position itself, literally overlooking the Chinese heartland, and with it the possibility of using the region as staging grounds for an invasion into the plains below. Even without India taking over the plateau, China is concerned that if it allowed political autonomy to Tibet, India could be permitted to establish military bases there.[87] But arguably even more important is a simple, but utterly vital resource that the Tibetan plateau provides: water.

Tibet is the highest and largest plateau in the world, and within its tens of thousands of glaciers it holds the largest store of glacial ice and permafrost outside the Arctic and Antarctic. This high plateau is often referred to as the planet's Third Pole.[88] The meltwater from these glaciers and snow forms the headwaters of ten of the largest rivers fanning out across the whole of South East Asia, including the Yellow River, Yangtze, Mekong, Indus, Brahmaputra and Salween. All these great rivers carry huge amounts of sediment eroded from the mountains to fertilise their flood plains and the rice paddies that have been established here.[89]

The Tibetan plateau thus serves as the water tower of the entire continental region, storing and distributing the precious resource along these rivers to provide drinking water, irrigation and hydroelectric power to more than 2 billion people.[90] It's this storehouse of vast amounts of freshwater, as well as the rich copper and iron ore deposits on the plateau,[91] that China seeks to control for its growing population and economy. By 2030, China is anticipated to have a 25 per cent shortfall in its water needs,[92] and so the Tibetan issue is no small matter. It's immaterial whether India would actually ever attempt to seize Tibet and restrict the flow of the rivers to turn off the taps on China's water – the mere possibility makes China vulnerable. Likewise, the concern from other downstream states such as India, Pakistan, Nepal, Burma, Cambodia and Vietnam

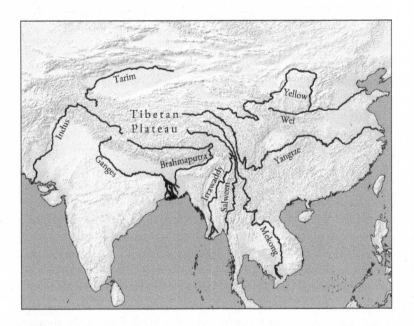

The major rivers radiating away from the third pole of the Earth, the Tibetan plateau.

is that in the future China could begin diverting the flow from these Tibetan rivers for its own internal usage.[93]

Irrespective of international criticism of China's occupation of Tibet, and human rights issues therein, these highlands represent an overpowering geopolitical concern for Beijing. It's for this reason that China is maintaining control, systematically constructing networks of road and rail links throughout the plateau and encouraging ethnic Han Chinese settlers into the area.[94]

Chapter 4

The Geography of the Seas

Oceans and seas cover nearly three-quarters of the Earth's surface. It's this fact that prompted the author Arthur C. Clarke to quip that we shouldn't call our planet Earth at all, but Ocean. And in terms of the themes of this book, the oceans are one of the best showcases for the close links between life on our world and deep space. The water on Earth is vital for all life, but when our planet formed from the swirling disc of dust and gas circling around the proto-Sun it was pretty dry. The Earth was too close to the Sun for there to have been much ice in the rocky material our planet coalesced from, and the heat of its formation melted the planet throughout and would have driven off water and any other volatile compounds. So the water that fills our oceans arrived after the Earth was born, brought by a bombardment of icy comets and asteroids from the colder, outer regions of the solar system – like a blizzard from deep space.

The oceans delivered by this extraterrestrial ice are of course enormously influential on the weather and climate systems of the planet, and water within the crust helps to lubricate the machinery of plate tectonics. But the oceans of the world are often considered to be just empty expanses. They are the blank spaces on our maps, the gaps on the page that merely define the outline of the landmasses. We have come to think that it is the continents and islands where history happens and upon which the human story has played out over the millennia. But the sea has its own rich story to tell.

TURNING WATER INTO WEALTH

From our earliest days, humanity has relied upon the planet's watery expanses for food. Fish taken from rivers, lakes or shallow coastal waters have provided easily accessible nutrition for tens of thousands of years.[1] But fishing in the open sea, far from land, requires much greater skills in shipbuilding and navigation. Norse seafarers were accomplished at long voyages and from around AD 800 had established an international trade in the dried cod they produced. These skills of the open sea were learned by other Europeans, and the North Sea became an important fishing ground.[2] And it's here that we can see how crucial the geography of the seas – and the landscape of the seafloor in particular – has been in history.

In the middle of the North Sea, between England and Denmark, lies Dogger Bank, a huge sandbank believed to be a large moraine that was piled up at the head of the Scandinavian ice sheet during the last glacial phase. During the lowered sea levels of the last ice age, this whole region would have been dry, known as Doggerland, and offered prime hunting grounds for our ancestors. Today it is submerged, but Dogger Bank forms a large area of shallow waters beneath the waves, and so provides a productive fishing area for cod and herring. ('Dogger' is an old Dutch word for a trawl-fishing boat.) Thus the Ice Age hunting grounds of our ancestors were drowned and transformed into a bountiful fishing region for medieval mariners.

This sandbank helped launch open-sea fishing in northern Europe from around AD 1000.[3] With increasing competition between fishermen and overexploitation of the closest shoals, Norse, Basque and other European seafarers were drawn further and further out into the North Atlantic in search of rich fishing grounds, first for cod and later for whales. European sailors ventured west, sailing past Iceland to Greenland, and then on to the north-east American coast where Norse fishermen established colonies on Newfoundland, half a millennium before Columbus set sail across the Atlantic. It was the lessons learned in the process – in seamanship and robust shipbuilding – that enabled European sailors to embark on the Age of Exploration in the early fifteenth century and build vast international trading empires (which we will ourselves explore in Chapter 8).[4]

But the same North Sea landscape has had another important influence on creating the modern world. The Low Countries of Belgium and the Netherlands sit on the flat coastline of the North European Plain, and from the thirteenth century the Dutch have been using windmills for drainage in order to create new farmland from the sea and marshes.[5] In effect, they are reclaiming portions of Ice Age Doggerland since it was re-submerged by the rising sea levels. But the building of dykes and windmills to reclaim tracts of land was expensive, and could only be financed with the pooling of resources from the community. The necessary funds were raised by the local church or council collecting loans from residents, and the agricultural profits from the newly reclaimed fields were then shared amongst those who had originally financed the project. Everyone in society soon came to invest their spare cash in the bonds sold to finance these large enterprises and this, in turn, created thriving credit markets. Shaped by the demands of its landscape, and the necessity to manage the sea, Holland became a land of capitalists.[6]

This system naturally transitioned into international commerce in the seventeenth century – it is a small step from buying shares in the construction of a local windmill to financing a trading ship bound for the Spice Islands. The practice of dividing the overall cost of a project into part shares also allowed investors to spread their risk – they could put small amounts of money into several voyages, so that if any one ship was lost, they would not be hit too heavily. This encouraged people to invest their money, rather than simply stashing it, which kept the interest rates on loans low and so the cost of capital cheap for further ventures. The Dutch also enthusiastically adopted and greatly refined the concept of the futures market. This is the ability to negotiate a price for a certain commodity at some point in the future – for example guaranteeing your price for 100 lb of cod landed from the Dogger Bank next week, or in a year's time. These derivatives can then be bought and sold themselves, like the actual products, creating a trade not of stock already in a storehouse but of abstractions.

The first national central bank as well as the first formal stock market were founded in Amsterdam in the opening years of the seventeenth century,[7] by which time Holland had become the most financially developed country in Europe.[8] These instruments of formalised capitalism quickly spread to other nations and created the financial institutions needed for the Industrial Revolution. Like the windmills in the medieval Netherlands, Britain's mills, factories and steam engines would have been prohibitively expensive to build without the pooling of capital from a number of different, and confident, investors.[9] Dutch financial innovations helped build the modern world, and they had grown out of her low-lying landscape and the need to reclaim land from the sea.

There are many other ways in which the salt waters of the planet have been instrumental in the human story. The sea can isolate a people from the rest of the word, which is what happened on Tasmania, for example. Here the inhabitants became cut off from the mainland with rising sea levels after the last ice age. The population on the island was too small to maintain technologies and tools such as fishing nets and spears, across the generations, and they became forgotten.[10] Or, as we have seen, the sea can help protect from invasion and maintain the independence of an island nation like Britain. The oceans are like the deserts on land: they are not inhabitable in their own right,* but they can be traversed for the

---

* In fact, as far as humans are concerned the seas of the Earth are barren deserts of water. As Samuel Taylor Coleridge wrote in *The Rime of the Ancient Mariner*: 'Water, water, every where, Nor any drop to drink.' Its salt levels mean that drinking seawater is lethal, and sailors need to carry stores of fresh water just like the caravans crossing the deserts.

movement of goods and people. Storm waves notwithstanding, the sea surface itself is a conveniently flat and unresisting medium for providing highways of trade across great distances. Ports are sited at the interface between sea and land, where goods are transferred from ships to river-boats or carts (or more recently, trains and trucks) to continue their journey to where they're needed inland, and many of these ports became prosperous and politically powerful cities. It was by mastering the navigation of the oceans that European states built vast maritime empires from the early sixteenth century onwards, projecting their power over immense distances with the help of fleets of cannon-sprouting, floating fortresses. And the choke-points in maritime routes, where ships are constrained to pass through narrow straits, are as strategically central to geopolitics and the power-play between states today as they were millennia ago.

In these different ways, the huge areas shaded blue on our world maps are just as important in shaping human history as the green, brown and white features indicating plains, forests, deserts and icy mountain ranges on land. Like this dry landscape the geography of the seas has directed our affairs through history. Let's start by looking at the Mediterranean.

THE INNER SEA

The Mediterranean region is one of the most complex tectonic environments on Earth. Here the African plate is shunting northwards and being subducted beneath the Eurasian, with a jumbled array of several smaller plates trapped in between, to drive a flurry of mountain-building and volcanic activity. The Mediterranean has also hosted a vibrant interplay of civilisations throughout history, with diverse cultures emerging, developing, trading resources and ideas, competing and going to war with one another, all within a relatively small and compact area. Could these two phenomena be related? Are there good reasons why the tectonic Mediterranean milieu offered a particularly fertile setting for nurturing ancient civilisations?

For millennia, the Mediterranean Sea has buzzed with maritime activity. From the Bronze Age merchants of the Minoans and Phoenicians, through the Greek city states and the Roman Empire, to the trade empires of Genoa and Venice in the later Middle Ages, this oval-shaped sea has connected peoples and cultures around its shores. The Mediterranean is an interior sea where passages are often just short hops. The high mountain ranges along the northern coast, created by the crunching tectonic plates, provide useful landmarks for navigation further from the coast. And the pinching narrowness of the Strait of Gibraltar, where it joins the Atlantic, means that generally the tides within the Mediterranean are minimal – a matter of mere centimetres – and there are no major surface currents to sweep you off course. The Mediterranean does experience fierce storms, however, and the wind patterns are complicated by air flowing off the surrounding landmasses. Still, on the whole this interior sea is ideally set up for communication and trade between cultures. Yet there has been a noticeable bias throughout history: the vast majority of civilisations have blossomed on the northern shores of the Mediterranean, and not on the southern.

Even if you just gaze idly at a map of the Mediterranean you'll notice something curious about the outline of the northern half compared to the southern side, the coast of Africa. The northern coastline is peppered with islands. These range in size from the tiny specks of the Cyclades archipelago in the southern Aegean to large landmasses a few hundred kilometres across – Sardinia, Crete and Cyprus. Many of these Mediterranean islands are now popular holiday resorts, but the sheer number of ancient ruins scattered across them attests to how critical they were in setting the stage for civilisation throughout antiquity. And it's not just the myriad islands poking above the Mediterranean waters that distinguish the north from the south. The coastline around the top lip of the Med is also fantastically detailed – it is full of inlets, coves, headlands and bays.

The shore and islands of the Aegean, the region that hosted many ancient Greek city states, for example, constitutes a full third of the length of the entire Mediterranean coastline but only a tiny fraction of its land area.[11] By stark comparison, the African shoreline is just a bit – well, plain. The coast running along modern Algeria, Tunisia, Libya and Egypt is monotonously smooth with essentially no offshore islands.

You may have thought that a land area broken up into many small parcels would be an encumbrance to early societies. But before the development of modern roads, railways and engines, travel and trade along overland routes was arduous. Transportation along calm rivers, or by sailing the seas, was much easier and faster, especially when carrying bulk cargo for long-distance trade. So the segmentation of the northern coast into many small pockets of land, separated by the relatively calm waters of the Mediterranean, was of huge assistance to the movement of people and goods between city states and kingdoms. The northern coastlines also provided a great selection of good natural harbours. In short, the northern Med is ideally set up for maritime activity and consequently many ancient cultures thrived along these northern shores.[12]

The African coastline forming the southern lip of the Mediterranean, on the other hand, is on the whole woefully unsupportive of seafaring societies. It offers very few protected natural harbours, and is backed by desert that has hampered agriculture and inhabitation. The cultures that did survive on the North African coast were generally restricted to thin slivers of land along the shoreline where farming was possible; but with the exception of the Egyptian civilisation supported by the mighty Nile, they could not extend very far inland. There have been some major ports on this African coastline of course. Carthage was located on the tip of modern Tunisia with a good natural harbour. This port began as a Phoenician colony in 814 BC and over the next five centuries came to dominate commerce in the western Mediterranean. It became a major rival to the Roman Republic, and the conflict between them led to a series of wars that resulted in the complete destruction of Carthage in 146 BC.*

Another major city on the North African coast was Alexandria, located in the delta of the River Nile. It was founded by Alexander the Great in 331 BC,[14] and on his death served as the capital of the Greek Ptolemaic dynasty for the next three centuries (until the death of Cleopatra in 30 BC). It also came to flourish as the major cultural and intellectual centre of the ancient world, not least because of its famous library. The city was built atop a stable bar alongside the vast river delta, and its 100-metre-high lighthouse tower on the island of Pharos guided ships into its port.[15] The location was carefully chosen. Alexandria was built on the western side of the Nile to prevent its harbour from silting up, as the water currents within the Mediterranean sweep the sediment pouring out from the river towards the east. This sediment is carried anticlockwise from the delta which smothers a huge area of the eastern Mediterranean, producing a straight, sandy coastline. It is not until as far north as Haifa, where a mountain juts into the sea to protect the bay beyond from silting up with the longshore drift, that the south-eastern Mediterranean offers a decent natural harbour.

So it was the dry climatic conditions of North Africa (to which we'll return in Chapter 7) and its unobliging coastline that conspired to hinder the rise of many great civilisations along this whole stretch of the Mediterranean. With the exceptions of Carthage and Alexandria, the roughly 4,000 kilometres of African coastline between the Strait of Gibraltar and the

---

* The Phoenicians were a civilisation very much born of their natural environment. Emerging around 1500 BC on the narrow but fertile strip of land that today forms the coastline of Syria, Lebanon and Israel, they had access to both natural harbours on the eastern Mediterranean coast and forests of cedarwood for shipbuilding.[13] The Phoenicians turned to the sea, and flourished for around a thousand years as expert mariners and merchants, establishing a widespread trade network and founding many colonies around the Mediterranean rim, including Carthage. They also invented the alphabet, and our word 'Bible' ultimately derives from the ancient Phoenician city of Byblos, which exported papyrus for writing.

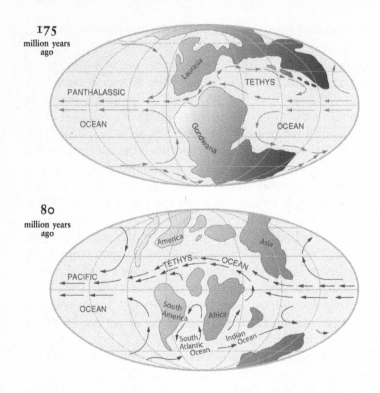

The closing of the Tethys Ocean to create the Mediterranean.

Nile Delta have been historically very quiet compared to the vigorous bubbling of different cultures, cities and civilisations around the northern shores.

But why is it that the northern and southern rims of the Mediterranean Sea, no more than a few hundred miles apart, are geologically so very different? Once again, there are planetary causes for this profound distinction.

Today's Mediterranean Sea is in fact nothing more than a puddle of water left behind after the disappearance of a once great ocean. About 250 million years ago, the face of our world would have been all but unrecognisable. The ceaseless roving of the planet's tectonic plates occasionally brings all the major chunks of

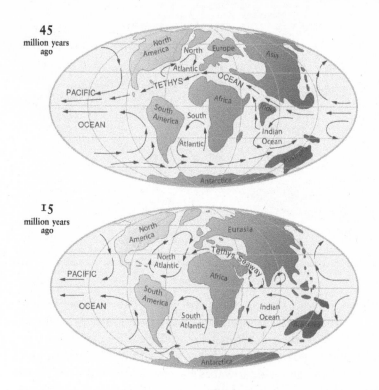

45
million years
ago

15
million years
ago

continental crust together to create a single, unified landmass: a supercontinent. And at the end of the Permian Period, the supercontinent Pangea – meaning the All-Land – stretched from pole to pole in a rough horseshoe shape, holding within its arms an ocean called the Tethys.* At this time you could have walked from the North to the South Pole across Pangea without ever getting your feet wet, although you would have needed to cross huge desert plains in the interior heartland of the immense continent.

But soon after the assemblage of Pangea had been completed, the supercontinent began to break up again. The land masses we are familiar with today tore from one another and rode into their current configuration. First, North America began

---

* The other side of the planet was covered by a vast uninterrupted ocean, Panthalassa, that was greater even than the Pacific today.

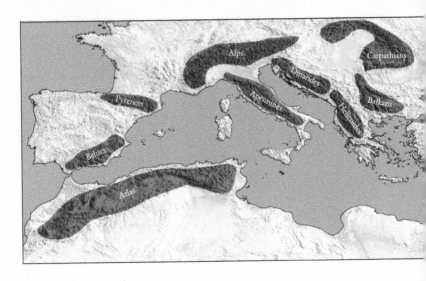

to tear away, unzipping along the seafloor-spreading rift that created the North Atlantic, and then South America ripped from Africa – their coastlines are still clearly complementary today. India separated from Antarctica and headed north, and Africa turned and rode towards Europe. In the last 60 million years, Africa, Arabia and India have all collided back into Eurasia, rippling up the great band of mountains along its southern margin from the Alps to the Himalayas.

Pangea is no more, and the Tethys Ocean has all but vanished. As the African plate slid northwards, the Tethys has steadily disappeared as its oceanic crust was swallowed beneath Europe, and its seafloor sediments crumpled into mountains. By around 15 million years ago the Tethys was no more than a narrow seaway, still open at both ends, between the North African coast and the Iberian Peninsula on one side, and through the Persian Gulf on the other; it also had a long northern arm inundating western Asia. But as the Red Sea rifted open it forced the Arabian Peninsula to swing away from the Horn of Africa and slam into the southern edge of the Eurasian plate, rippling up the Zagros mountain range. This created the Middle East region as we know it today, and closed off the eastern opening of the Mediterranean. The northern

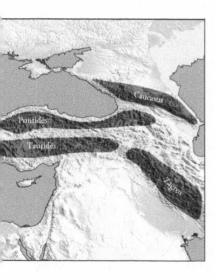

The Mediterranean today, rimmed by mountain ranges thrown up by the closing of the Tethys Ocean.

arm of the Tethys dried up, leaving behind as remnants the Black, Caspian and Aral Seas across western Asia. Meanwhile, as Africa was still pushing north, its north-western tip crunched into the Iberian Peninsula, finally cutting off the Mediterranean from the Atlantic on its western end between 5.5 and 6 million years ago.[16]

Now completely severed from the rest of the world's oceans, and lying in a hot climatic zone, the Mediterranean was losing water to evaporation faster than the rivers flowing into its basin were able to top it up, and it rapidly dried out. As the water level dropped, the Mediterranean was split into two halves by a ridge that continues as an off-shoot of the Atlas Mountains in Tunisia.* The western half of the Mediterranean desiccated completely, and laid down great salt deposits on the sun-baked floor. In fact, the sheer thickness of these deposits – up to 2 kilometres in places[17] – beneath the Mediterranean today indicates that the sea must have dried out and been refilled by the Atlantic spilling back in many times in succession.[18] This process reduced the salt content of the world's oceans by about 6 per cent.[19] The eastern

---

* The port of Carthage lay on this raised lip, and the island of Sicily and the 'toe' of Italy are peaks of this same barrier.

Mediterranean basin is deeper, and received some flow from the Nile and the Black Sea via the Bosphorus, so although its waters dropped thousands of feet below sea level it didn't completely dry out but persisted as a brackish lake, not unlike the Dead Sea today.

Then, around 5.3 million years ago, ongoing tectonic activity saw the western rim of the basin subside again and the Mediterranean permanently reopened. Waters of the Atlantic began trickling in, before developing into a surging torrent with huge volumes of water gushing down the slope to refill the empty, dusty Mediterranean basin, perhaps in as little as two years. The current Strait of Gibraltar was gouged out by this scouring megaflood.[20]

The Mediterranean is still shrinking today as the African tectonic plate continues its northward march, and will eventually disappear completely. And it is this tectonic process that explains the geological differences between the sea's northern and southern coastlines. The southern Mediterranean coastline is relatively smooth and bereft of natural harbours because the African plate is being tipped downwards to be subducted and destroyed beneath the Eurasian plate. The entire northern Mediterranean coastline, on the other hand, is mountainous because of this continental collision. Here the combination of tectonic subsidence and the fact that we are currently in an interglacial period with high sea levels has produced a submergent coastline. The fantastically intricate northern Med with its multitude of islands, headlands, bays and an abundance of protected, natural harbours is the result of this drowned landscape. It is this fundamental tectonic fact that has empowered seafaring cultures along the northern rim, and so influenced history from the Bronze Age to the present day.

SINBAD'S WORLD

The interior sea of the Mediterranean linked together the cultures at the far western extremity of the Eurasian continent into a great trade network. But maritime trade over much greater distances has also shaped the history of civilisations. Through the ages, numerous cultures and empires have emerged across the southern half of Eurasia, the region to the south of the great band of arid steppe grasslands ranging across the continent to which we'll return in Chapter 7. These societies traded with each other by sea routes along the southern margin of this broad continent.

The maritime routes linking eastern and western Asia reached across the Indian Ocean. By around 3000 BC, merchants in Mesopotamia were transporting their goods south to where the Tigris and Euphrates rivers merge and flow into the Persian Gulf. From here, they sailed down the Gulf, through the narrow Strait of Hormuz at its mouth, and then along the South Asian coast to the mouth of the Indus river. As civilisation spread to Egypt, Phoenicia and Greece on the Mediterranean shores, a second major artery of trade opened. From the Nile Delta, goods were carried overland by camel caravans across the mountainous Eastern Desert to ports on the Red Sea. From here, ships sailed down the long channel of the Red Sea, around the southern edge of Arabia, and then entered the Indian Ocean.[21]

It was not an easy journey. Hidden shoals along the Red Sea coast made navigation potentially treacherous, the heat was punishing, and the extreme aridity of the region, with vast deserts on either side, meant there were few sources of fresh water along the shores. Indeed, the narrow strait forming the entrance into the Red Sea came to be known by Arab sailors as Bab-el-Mandeb – the 'Gates of Woe'. Before embarking on the long passage up the Red Sea, ships would call into the port of Aden, sitting just around the lip of the Arabian Peninsula and commanding the Bab-el-Mandeb gateway. Nestled in the crater of an extinct volcano, Aden was a vital stop for taking on water, and as a busy entrepôt developed into a prosperous and well-fortified city.*

* Aden was also of key strategic value to the British from the mid nineteenth century. The port lies roughly equidistant from the Suez Canal, Mumbai on the western Indian coast and Zanzibar in East Africa, all of which were under British control at the time. During the heyday of steamships, Aden was an important way-station for loading coal and boiler water. It was for exactly the same reasons that the US annexed Hawaii in 1898, which served as a coaling station for American naval operations in the Pacific.[22]

Both the Red Sea and Persian Gulf routes into the Indian Ocean buzzed with merchant shipping and both these maritime thoroughfares are consequences of the same episode of tectonic activity.[23] As we saw in Chapter 1, the Red Sea is one of the three branches of the Y-shaped system of rifts

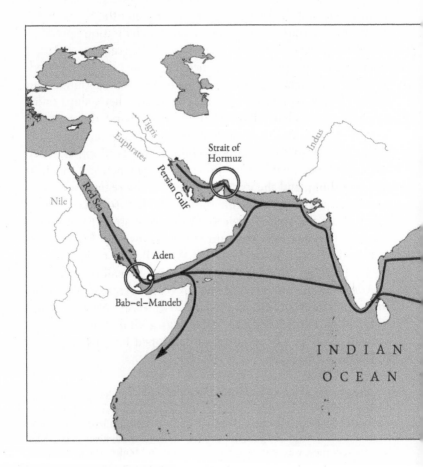

that ripped open the skin of the planet as a huge magma plume swelled beneath the African crust. The growth of the southerly branch, the East African Rift, set the stage for our evolution as a species, while the deeper fracture to the northwest tore off the Arabian Peninsula as a shard from Africa,

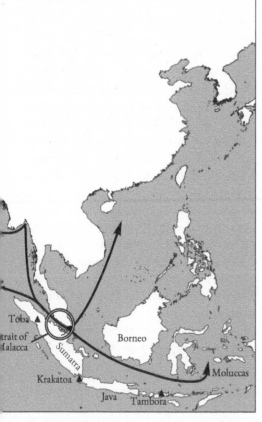

The major east–west Eurasian maritime trade routes and crucial straits.

with water pooling into this 2,000-kilometre-long crack to create the Red Sea.*

The Arabian Peninsula remains hanging off Africa by only a narrow sinew of land in the north – the Sinai Desert – and as the Red Sea has widened the Arabian block has swung east to slam into the southern edge of the Eurasian plate. This folded up the Zagros Mountains in Iran, and along the foot of this range, where the crust has been depressed down into a wedge-shaped foreland basin, the Indian Ocean washed in to create the Persian Gulf.

The earliest trade routes from the Red Sea and the Persian Gulf to India hugged the coastline. But by around 100 BC[24] the merchants of Ptolemaic Egypt had discovered how to use the south-westerly monsoon winds in summer to sail from Bab-el-Mandeb directly across the Indian Ocean to the west coast of India in just a few weeks,[25] returning in winter when the monsoon winds reversed direction. Exploiting this feature of the planet's atmospheric patterns – to which we'll return in Chapter 8 – led to a surge in maritime commerce across Eurasia.[26] But by the end of the seventh century AD, the Islamic conquests across Arabia, North Africa and south-western Asia had closed the gates of Bab-el-Mandeb to European sailors. For centuries the dhows and caravans of Muslim merchants now dominated the three great east–west trade routes across Asia: the maritime passages from the Red Sea and the Persian Gulf across the Indian Ocean, and the Silk Road through Central Asia.[27] This is the world of Sinbad the Sailor in *One Thousand and One Nights*, who loaded up with trade goods in Baghdad and set sail from Basra down the Persian Gulf on his seven adventurous voyages.

Prior to the rise of Islamic supremacy over these trade routes, India had been well known to Greek and Roman geographers

---

* Further fracturing of the continental crust at the northern end of the Red Sea formed the narrow gulfs of Suez and Aqaba, the latter of these splits extending to form Lake Galilee, the Jordan valley and the Dead Sea, whose shores are, at 400 metres below sea level, the lowest-lying land on the Earth's surface.

like Strabo and Ptolemy, but after the blocking of the Red Sea passage knowledge of its location faded into the obscurity of myths.[28] It would be the best part of another millennium before Europeans once again sailed into the Indian Ocean, as we'll see in Chapter 8. And when they did, they would find a trade network in South East Asia just as vibrant as that of the Mediterranean.

SPICE WORLD

Indeed, in many respects the maritime area of South East Asia is much like the Mediterranean. But rather than an internal sea bounded on all sides by land, this region is a dispersed splattering of islands open on both sides to the expansive Indian and Pacific Oceans. The East Indies are part of the continental shelf of Eurasia: here the seas are relatively shallow and the land masses are simply the higher ground of this landscape poking above the waves. Like the northern rim of the Mediterranean, the margins of this region are volcanically active, as the Indo-Australian and Pacific plates are being subducted under the Eurasian to melt and release rising blobs of magma.

A whole chain of volcanoes runs along the backbone of Sumatra and Java, and curves all the way round to the Banda Islands. This volcanism has produced fertile soils, but also some of the most violent eruptions in history, such as Tambora in 1815 and Krakatoa in 1883. The eruption of the Toba supervolcano in Indonesia around 74,000 years ago was the largest of the past two million years. It ejected an enormous amount of ash that smothered 1 per cent of the planet's surface and may have darkened the skies sufficiently to cause a global chilling for several decades. (This has even prompted the controversial claim that the Toba eruption caused a crash in the surviving population of humanity.)[29]

Whereas the Mediterranean sports a few hundred islands, South East Asia contains over 26,000, ranging from

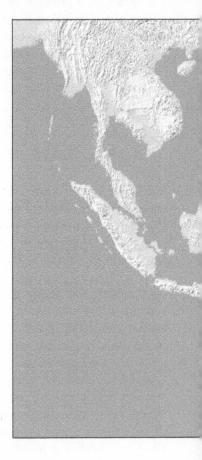

The South East Asian archipelagoes, and the tiny size of the spice islands of the Moluccas and Bandas.

thousand-kilometre-long landmasses like Borneo and Sumatra to minute specks of calderas. This extreme dispersal of the land area, coupled with the rugged, mountainous terrain of the islands, inhibited the unification of territory into large empires as had happened in China or around the Mediterranean.[30] Yet trade flourished across these South East Asian seas. Alongside cotton from India, porcelain, silk and tea from China, and precious metals from Japan,[31] the most highly valued commodities were spices: pepper and ginger

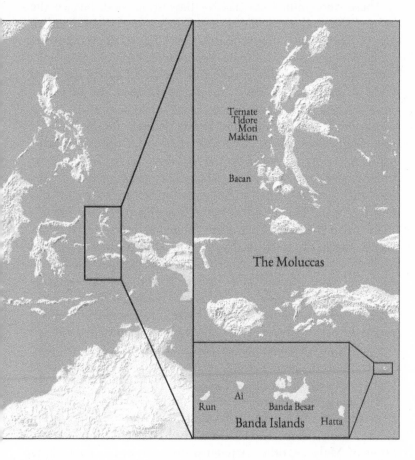

Ternate
Tidore
Moti
Makian

Bacan

The Moluccas

Ai

Run   Banda Besar

Banda Islands   Hatta

from India, cinnamon from the island of Ceylon (Sri Lanka), and nutmeg, mace and cloves from the 'Spice Islands' of the Moluccas.*

---

* India's black peppercorns are botanically very different from bell (sweet) peppers and chilli peppers, which are both fruits of Capsicum plants native to Central and South America. These New World species were unknown to the rest of the world until the great fifteenth-century transfer of domesticated plants and animals that occurred after the European discovery of the Americas, known as the Columbian Exchange.

These were valued not just for flavouring food, but for the aphrodisiac and medicinal properties they were considered to have.[32] The spices came from different kinds of plants growing in the region's tropical climate. Pepper is the fruit of a rainforest vine, ginger a root, cinnamon a tree bark, and cloves the dried buds of unopened flowers. Nutmeg is the seed and mace the seed covering of the same evergreen tree.[33] Some of these plants were widespread through the region. Pepper, for example, is found right across South and South East Asia, although historically most was produced on the Malabar Coast of south-west India.[34] Here the Western Ghats, a low range of mountains, trap the rainfall from the summer monsoons to produce a moist, tropical climate ideally suited to this particular vine.[35]

But other spices were extremely limited in their native habitat. Cloves originally grew in the volcanic soils of only a handful of small islands in the northern Moluccas archipelago: Bacan, Makian, Moti, Tidore and Ternate.[36] And the nutmeg tree appeared on only nine pinprick-sized islands – the Bandas – further south in the Moluccas.[37] These rare spices commanded premium prices, especially by the time merchants had carried them all the way west to the Mediterranean: the commercial importance of these minute specks of volcanic islands was vastly out of proportion to their size.[†]

The maritime trade network of South East Asia was far larger than that of the comparative puddle of the Mediterranean. Routes from the Indian Ocean threaded through the narrow Strait of Malacca, others stretched down from the East China Sea, and also from the Moluccan Spice Islands in the east, all converging on trading ports on the Malay Peninsula or the islands of Java and Sumatra.[38] By AD 1400, the port of Malacca

[†] So much so that in the late seventeenth century, after the Second Anglo-Dutch War, it was agreed that the Dutch claim over Manhattan be ceded to the English in exchange for the spice island of Run, one of the smallest Banda islands. Run is just 3.5 kilometres long, but its acquisition allowed the Dutch to secure their nutmeg trading monopoly in the East Indies. Manhattan was swapped for nutmeg – and New Amsterdam renamed New York.

on the south-east Malaysian peninsula had grown from a small fishing village to one of the largest centres of maritime trade in the world.[39] It was strategically located roughly halfway along the 800-kilometre Strait of Malacca between the Malay Peninsula and the long island of Sumatra, at a point where the funnel-shaped strait narrows to just 60 kilometres wide. The Strait of Malacca was one of the most important waterways in the Eastern Hemisphere as it served as the crucial marine thorough-fare between the Indian Ocean and South China Sea.[40] The port's heaving markets burst with a huge diversity of trade goods: wool and glass from Venice, opium and incense from Arabia, porcelain and silks from China, and of course the spices from the Bandas and the Moluccas.[41] Malacca was one of the most cosmopolitan places on Earth, its port a forest of masts where the dhows from the Indian Ocean berthed alongside the junks from China and the Spice Islands, and with a population larger than Lisbon speaking scores of different languages that could be heard over the din of the markets.[42] It was the riches of this spice trade that provided the major draw for European navigators trying to find new sea routes to the east at the end of the fifteenth century.*

And when they arrived, they strove to dominate this expansive South East Asian trade network by capturing key features of the geography of the seas: naval chokepoints. But to illustrate the histor-ical importance of these features we'll turn first to ancient Greece.

BOTTLENECKS

As we saw earlier, the rugged landscape of Greece makes for a coastline with many inlets, bays and channels for natural

---

* Europe at the time did already have access to many herbs and spices – saffron grown in Spain after its introduction by Arab traders, coriander and cumin native to the eastern Mediterranean, as well as the aromatic species indigenous across Europe: rosemary, thyme, oregano, marjoram and bay. But the exotic pepper, nutmeg, mace and cloves from the East were much rarer and therefore valuable in western markets.[43]

harbours and so nurtured vigorous trade by sea. In fact, this mountainous geography is believed to have been influential in maintaining the Ancient Greek city states as autonomous entities. Steep-sided ridges running down to the coastline physically separated them from each other, and prevented any one state gaining complete dominance to build an empire. The result was a world of many independent city states both sharing a common culture and language and competing with each other in a constantly shifting pattern of allegiances and conflicts.[44†] But at the same time, the paucity of coastal plains restricted the area available for productive agriculture. Unlike Mesopotamia or Egypt, Greece is not blessed with alluvial plains of deep, rich soil, and whilst there are fertile valleys inland, there's not a great many of them. Greece's mountainous terrain generally offers only a thin, light soil, which is mostly dry because of scant and unreliable rainfall, and there are too few large rivers to make widespread irrigation possible. Indeed, except for the Rhone on the western extremity of the Alps, the major rivers of Europe are blocked from flowing towards the Mediterranean by the chains of mountains thrown up by the continental collision.

These combined environmental factors mean that historically the peninsula has struggled to grow adequate grain as a staple to feed its population, and many of the Greek city states lived

---

† The landscape also dictated the nature of Greek warfare. The rugged terrain of narrow gorges and steep mountains and hills is not conducive to the wheeling chariot battles that had been common across the plains of Asia; nor is it well suited to cavalry formations. Instead, the Greek states developed armies of hoplites, foot soldiers armed with spear and shield who by the seventh century BC were trained to fight in tight phalanx formations. These hoplite armies were made up not of professional soldiers but citizens – farmers, craftsmen and traders – who brought their own bronze weapons and armour. Thus Greek battles weren't decided by an elite class riding chariots or on horseback, but by the common citizens working together, each trusting the man to his right in the phalanx to protect him with his shield. This solidarity among the free men in Greek culture contributed to the early development of democracy within some city states, in particular Athens (although women, slaves and those who were not landowners were still excluded from the political process).[45]

with the constant threat of food shortage and famine. The Greek climate, however, is well suited for producing olive oil and wine, as well as for rearing flocks of goats and sheep, all of which could be traded for wheat and barley grown overseas.[46]

Around the same time that some of the Greek city states were developing the first democracies in the world, early in the first millennium BC, their populations began to outgrow the home-grown food supply provided by their surrounding environment. Thus the Greeks looked to other lands around the Mediterranean to source the vital grain they needed to feed themselves. Sparta, Corinth, Megara and their allies sent their ships west to bring back grain. Sicily was colonised to reap the benefits of the rich, volcanic soils around Mt Etna.* A second set of allied Greek city states around the Aegean Sea, including the flourishing city of Athens, established colonies in the fabulously fertile valleys of the Dnieper and Bug rivers along the north shores of the Black Sea, the westernmost extension of the Eurasian steppe grasslands (we'll return to this region in Chapter 7).[47] To get there, Greek ships had to sail through two exceedingly narrow straits between the Aegean and the Black Sea: first they had to negotiate the Hellespont, or the 'Bridge of the Greeks' (now called the Dardanelles) into the small Sea of Marmara, and then the even tighter Bosphorus into the Black Sea.†

As Greece grew ever more populous, fed on grain imported from its overseas breadbaskets, the rivalry between the two alliances of city states, led by Athens on one side and Sparta on the other, grew increasingly fractious. Eventually, in 431 BC this erupted into the devastating Peloponnesian War. It ground on for almost thirty years, with both sides attempting

---

* Mt Etna is the tallest active volcano in Europe and one of the most active volcanoes in the world, erupting regularly with magma generated as the African plate is subducted beneath the Eurasian.

† The Dardanelles are not just a vital marine chokepoint between the Mediterranean and Black seas, but also a strategic crossing point from Europe into Asia Minor. Alexander the Great crossed eastwards here in 334 BC to conquer Persia.[48]

to take control of the seas, but ultimately it was Athens' dependence on grain imported along the sea lanes from the Black Sea that proved her fatal weakness. The Spartans realised that they didn't need to attack Athens directly but merely sever her lifeline. They gathered their naval forces in 405 BC and waited until midsummer to strike, when the greatest number of Athenian grain ships were preparing to sail from the Black Sea with their precious cargo before autumn drew in and closed the shipping route with its stormy seas and overcast skies.* Falling upon the Athenian navy at the Battle of Aegospotami, right in the narrow strait of the Hellespont, the Spartans utterly destroyed it – over 150 ships were sunk or seized. Having taken control of this vital chokepoint on the sea route from the Black Sea, the Spartans did not even attempt a final assault on Athens: they knew the cold spear of starvation would be far more devastatingly effective than those wielded by its hoplite army. Athens had no option but to sue for peace under humbling terms, losing the rest of her fleet and her overseas territories.

The Peloponnesian War is a good illustration of the central importance of the geography of the seas and the vulnerability of vital maritime routes at narrow straits. Commanding such naval chokepoints, and thus a rival's access to overseas resources, is often as important as controlling territory on land, and can determine the outcome of wars and the fate of civilisations. Alongside the bottlenecks at the Dardanelles and the Bosphorus, the Strait of Gibraltar – the thin tongue of sea between the Iberian peninsula and the Tangier coast – has played an important role in controlling the naval traffic between the Mediterranean and Atlantic, and provided the setting for the Battle of Trafalgar in 1805 between the Royal Navy and the combined fleets of the French and the Spanish.

---

* Before the invention of the magnetic compass for navigation, sailing across open sea was just too dangerous when the night-time stars were obscured from view.

Other straits around the globe have proved as critical to world history. When European sailors reached the Indian Ocean from the early fifteenth century – first the Portuguese, and then the Spanish, Dutch and British – they attempted to command a whole set of chokepoints to exert control over an entire region of the Earth's oceanic surface.

As we saw earlier, for millennia two main sea routes carried trade between Egypt and the Middle East, and India: the passage along the Red Sea and that down the Persian Gulf. Both are linked to the open Indian Ocean through the narrow straits of Bab-el-Mandeb and Hormuz. And from India, the trade route to the major entrepôt ports on the islands of the East Indies threaded through the Strait of Malacca. For the merchants who had sailed around South East Asia for centuries the seas were an open commons, a vast region of free trade for all. Duties were levied at ports, and pirates were a rumbling concern, but no navies harassed foreign ships on the open seas. But the Europeans had a very different mindset, born of their heritage of naval warfare around the Mediterranean and North Atlantic. These colonial powers were bent on dominating the trade networks to establish a monopoly for themselves. To achieve this, they built fortresses to protect key ports and patrolled the waters with their warships to aggressively suppress competitors. Most crucially, they attempted to capture the naval chokepoints of Bab-el-Mandeb, Hormuz and Malacca to seal off the maritime routes to all but their own shipping, controlling trade across the whole reach of the Indian Ocean by commanding just a few critical locations in the geography of the seas.*

And naval chokepoints are just as strategically crucial today. It is no longer the spice trade that gives them acute geopolitical importance but the transport of another resource of global

* When in 1611 the Dutch established a new, faster passage from South Africa to the East Indies – the Brouwer Route which we'll discuss in Chapter 8 – the key gateway, and thus their strategic focus, shifted from the Malacca Strait to the Sunda Strait between Java and Sumatra.

significance. Oil today occupies almost half of the total shipping tonnage around the world,[49] and its continued unrestricted flow is critical to the current global economy.

BLACK ARTERIES

Oil not only fuels our modern world, but lubricates machinery, coats our roads, provides plastic and pharmaceuticals, and is used in the production of artificial fertilisers, pesticides and herbicides that help produce the food we need. Over half of the global oil supply is delivered by tankers moving along the worldwide network of sea lanes,[50] and thus passing through natural straits. As we have seen, the straits of the Dardanelles (or Hellespont) and the Bosphorus have been strategically critical since the time of the Peloponnesian War. Ukrainian grain is still exported across the Black Sea, but now around 2.5 million barrels of oil are also carried every day by tankers through this pair of Turkish straits to supply Southern and Western Europe with fossil fuel from Russia and the Caspian Sea region. The Bosphorus, less than a kilometre wide, is the narrowest strait being navigated by major vessels in the world.[51]

We have also constructed artificial chokepoints with our canals that link seas to create more direct shipping routes, such as the Panama Canal and the Suez Canal. When the Suez Crisis in 1956 closed the canal for six months and forced shipping to re-route around southern Africa, the result was fuel shortages across Europe.[52] Yet by far the most strategically critical strait in this present age of oil is Hormuz.[53]

We'll see in Chapter 9 how oil was created by our planet, and why so much of it can be found in the Middle East. The Persian Gulf produces about a third of the global oil supply[54] and Iraq, Kuwait, Bahrain, Qatar and the United Arab Emirates must all ship their oil exports through the Strait of Hormuz; only Saudi Arabia and Iran are able to make use of alternative maritime links to the ocean shipping lanes. As a result the strait is frenetically busy with tanker traffic, transporting 19 million barrels every day – one-fifth of the world's supply.[55] But it also means that this artery carrying the black blood to fuel the world is exceedingly vulnerable as it passes through the straits. It has been calculated that over the forty years since the Arab oil embargo of 1973, the United States has spent over $7 trillion on its military presence in the Gulf to secure the steady flow of oil for the global markets.[56] While piracy and terrorist attacks are concerns, the greatest fear is that international relations with a state like Iran might sour to such a point that they slam shut this vital chokepoint and put a stranglehold on world oil supplies.[57]

About 10 per cent of the oil produced around the Persian Gulf is shipped round the Cape of Good Hope to the United States, with a smaller portion being sent through Bab-el-Mandeb, up the Red Sea, and through the Suez Canal to the Mediterranean. But the lion's share travels the millennia-old shipping route round India to East Asia, threading through the narrow bottleneck of the Strait of Malacca. About a quarter of all oil transported by sea – roughly 16 million barrels a day – passes through this strait in tankers, and then on to feed the economies of China and Japan, as well as South Korea, Indonesia and Australia.[58]

While the nature of the major commodities may have changed through history – from grain to spices to oil – the role played by the geography of the seas and the strategic importance of naval chokepoints has remained ever critical. Before the advent of railways, automobiles and air travel it was the seas that facilitated long-distance trade. Even today 90 per cent of world trade is still carried by shipping.

But the role of the oceans consists of more than just providing maritime highways for long-distance trade and chokepoints that define much of the current geopolitical landscape. Let's explore now how the geography of the seas can also shape the economics and politics of a nation.

BLACK BELT

When the American colonies declared their independence from British rule in 1776, and fought and won the Revolutionary War, their population was still almost entirely huddled along the Eastern Seaboard. Over the following decades the United States underwent a prodigious expansion, encouraging settlers to head west and acquiring huge areas of territory in a series of purchases and annexations. Within a century of its birth as a nation, the United States had quadrupled in size, and stretched from sea to shining sea across the entire breadth of the continental landmass. The USA had become a de facto island nation, shielded on the east and west by the Atlantic and Pacific Oceans, whilst also enjoying access to maritime trade with Europe on one side and Asia on the other. America was able to accomplish economic success and champion ideals of liberty precisely because of this safety from external threats, born of its geographical circumstances. While European nations continued to jostle with each other on their crowded continent, America's territorial security engendered an isolationist stance in its foreign policy for almost two centuries.*

But there's another way that the sea has left its imprint on American politics, with roots reaching much further back in our planet's history.

In the November 2016 US elections the Republican nominee Donald Trump beat his Democrat rival Hillary Clinton to

---

* The island nation of Japan also underwent over two centuries of isolationism from the 1630s. During the Edo period, the *sakoku* ('closed country') policy barred most foreigners from entering Japan, and prohibited the Japanese from travelling overseas or building ocean-going ships. The only connection with the outside world was through a single trading post the Dutch were permitted to run on a tiny island within the bay of Nagasaki. Diplomatic and trade contact was reestablished after 1853 when steam-powered American warships arrived at the Japanese capital and forced the government to open their nation to the world.

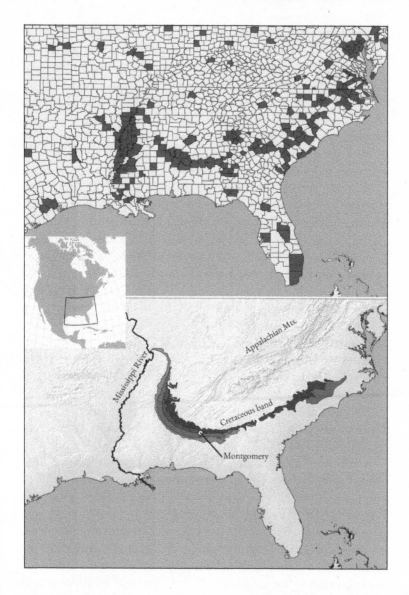

The pattern of Democrat-voting counties (dark) within the sea of Republicans in the south-eastern US (*top*) follows the Mississipi and the arc of 75-million-year-old Cretaceous rocks (*bottom*).

become the 45th president of the United States. A map of the results shows the blue Democrat-voting states of the north-east and up the Western Seaboard, along with Colorado, New Mexico, Minnesota and Illinois, whereas huge areas in the centre of the country are coloured Republican red. The states in the south-east also voted Republican overall, including Florida, which tipped to the Republicans in this election. But a look at a finer-resolution map of the voting behaviour, showing individual counties, reveals something very curious.

Running through the wide expanse of red in the south-east, there is a very distinct blue line of counties that voted strongly Democrat, curving through North and South Carolina, Georgia, Alabama, and then down the banks of the Mississippi river. And this blue ridge isn't just a quirk of the most recent presidential election. It is also apparent in the 2008 and 2012 elections won by the Democrats under Barack Obama, as well as the preceding terms of George W. Bush. In fact, this voting feature recurs through time right back to the reconstruction of the United States after the Civil War. What could be the underlying cause of this pattern in the south-eastern states that has endured for so long in something as variable and fluid as presidential politics and elections through history?

The astonishing fact is that this clearly defined band of Democrat-voting areas is the result of an ancient ocean, tens of millions of years old.

If you look at a geological map of the US you'll notice that the pattern of blue counties follows along a curved band of surface rocks that were laid down during the late Cretaceous Period in Earth's history, between 86 and 66 million years ago. This relatively narrow band of exposed Cretaceous rocks curves around older rocks further inland to the north, including the high relief of the Appalachian Mountains, and disappears underground to the south where it is overlaid by more recent rock deposits.

During the Cretaceous Period, when the climate was hot and sea levels far higher than they are now, much of what is today the United States was flooded. The sea protruded right up through the middle of the US as the Western Interior Seaway, and lapped at the toes of the Appalachian Mountains along the eastern side of the continent. Material eroded out of the Appalachians and carried by rivers into this shallow sea was deposited as clays on the seabed. In time, these seafloor clays turned into a layer of shale rock. As sea levels fell again, the outline of the US that we would recognise today emerged, and erosion re-exposed a strip of these ancient seafloor sediments on the coastal plain. Soils derived from this band of shale bedrock are dark and rich in nutrients originally eroded from the mountains. The term 'Black Belt' initially referred to this stripe of distinctively coloured and agriculturally productive soils through Alabama and Mississippi.

These dark, rich soils derived from the Cretaceous shale were perfect for growing crops, and in particular the cultivation of cotton. With the Industrial Revolution gathering pace and accelerating the processing of cotton into clothing – brought about by the mechanisation that rapidly separated the cotton fibres from their seeds, spinning them into thread and then weaving finished textiles – demand for cotton surged, making it a key cash crop. But the cultivation of cotton was deeply labour-intensive. Unlike cereal crops where the desired grain can be simply shaken from the plant stalks by a threshing machine, early cotton cultivation needed nimble human fingers to pluck each fluffy boll individually from the shrub. And from the late eighteenth century in the southern states these were provided by slaves.

By 1830, slavery had become solidly established in South Carolina and along the Mississippi, and by 1860 it had spread up from the Gulf coast of Alabama and through Georgia. By the height of slave-tended cotton plantations, the term 'Black Belt' had come to take on a different meaning, describing the populations found in the Deep South – a dense concentration of African-Americans along the banks of the Mississippi and around the curve of the Cretaceous rock belt underlying the soils.[59]

Even after the Confederacy lost the American Civil War in 1865 and slavery was abolished across the southern states, there was no sudden change in the demographics or indeed in the economic focus of this region. The former slaves continued to work on the same cotton plantations, but now they were sharecropping as freedmen. But the economic fortunes of the Deep South began to slump with falling cotton prices followed by the infestation of the cotton-growing areas with the boll weevil in the 1920s. Several million African-Americans migrated from rural areas in the southern states to the major industrial cities of the north-eastern and midwestern United States, especially after the Great Depression of the 1930s. Yet the largest populations of African-Americans remained in the regions where they had the greatest initial density: the historical 'Black Belt' of fertile soils.

After the Second World War the 'Black Belt' therefore formed the heartlands of the civil rights movement. Rosa Parks refused to give up her bus seat to a white traveller in December 1955 in the city of Montgomery, Alabama, smack in the middle of this curving strip of 75-million-year-old Cretaceous rocks. Even today, virtually all the counties in the US with the highest proportion of African-Americans lie along this same arc within the south-east.[60] Persisting after many African-Americans had migrated north and west, these populations are almost like an erosional remnant staying in place after the economic tide has swept millions elsewhere.

Without major development from industry or tourism, this formerly economically productive region has long suffered socio-economic problems of high unemployment and poverty, low levels of education and poor health care. Thus the electorate here has traditionally tended to vote for the policies and prom-ises of the Democrat Party, producing the distinct curving band of blue in the presidential election maps. There is a clear causal chain taking us from the politics and socio-economic conditions of today, to their roots in historical agricultural systems, and then further back to the geological tapestry of the ground beneath our feet. The exposed band of ancient sea-floor mud is still imprinted on our political map.

Chapter 5

What We Build With

Who built the pyramids?

Your immediate answer may well be the pharaohs of ancient Egypt. And you would be right, of course. It was the all-powerful god-kings of the fertile Nile Valley who, over 4,500 years ago, were able to muster and orchestrate the manpower needed to quarry, transport and assemble giant stone blocks into the colossal pyramids towering over the Giza Plateau. The largest of these is the Great Pyramid, built during the reign of the pharaoh Khufu – or Cheops, as he is also known – and completed around 2560 BC. Until the completion of Lincoln Cathedral in 1311 it was the tallest human-made structure in the world.[1]

The main bulk of the Great Pyramid is made up of about 2.5 million limestone blocks, each weighing 2.5 tonnes on average, laid atop each other in 210 layers. These were quarried from a limestone deposit nearby, dragged on sledges to the construction site and then hauled to the top of the growing pyramid up earthen ramps. This pointed construction was then covered in outer casing stones, using a much higher-quality limestone quarried further away on the other side of the River Nile, which were fitted together tightly and then beautifully polished. The Great Pyramid would originally have gleamed spectacularly in the sun, but most of these casing stones have since been removed. The large granite blocks, some weighing up to 80 tonnes, used for lining the interior chambers were quarried much further away in Aswan, about 400 miles upriver.

It is thought that the construction of the Great Pyramid took several decades and a workforce of tens of thousands of skilled labourers, who were paid in bread and beer. They worked without iron tools, pulleys or wheels, instead using copper chisels, drills and saws. But although the scale of the Great Pyramid is absolutely staggering, and the human effort involved in its construction was truly colossal, what is perhaps equally amazing is the nature of the building materials. It turns out that they were created by some of the simplest organisms on the planet.[2]

BIOLOGICAL ROCK

If you can get up close to the massive building blocks making up the core of the Great Pyramid – now exposed with the removal of the outer casing stones – and peer intently at their surface, you'll notice a very curious texture. The limestone blocks are made up of scores of coin-like disks. Search for some that have cracked open and you might be lucky to see their internal structure: an impressively intricate spiral, subdivided into small chambers. You are looking at fossils of sea creatures called foraminifera, or forams. And most impressive of all, given that each shell can be up to a few centimetres across, is the fact that the organism that created them is single-celled. The largest human cell, a woman's egg cell, is about a tenth of a millimetre across, and only just visible to the unaided eye. The sea creatures that make up the limestone of the pyramids are absolutely colossal in comparison. They belong to a kind of giant foram called *Nummulites* (meaning 'little coins' in Latin).

Deposits of nummilitic limestone are found not just around the Nile, where they offered construction material for the ancient pyramid-builders, but across a huge area ranging from Northern Europe to North Africa, from the Middle East to South East Asia. This expansive region of nummulitic limestone was laid down in the warm, shallow margins of the Tethys Ocean, 40–50 million years ago. During this period of the early Eocene, global temperatures were elevated for longer than the PETM extreme temperature spike that we explored in Chapter 3, although they weren't as high. The high sea levels caused the Tethys Ocean to flood in great arms stretching to Northern Europe and across North Africa. Huge numbers of forams lived in the warm waters, and when they died great drifts of their coin-shaped shells made of calcium carbonate sank down and carpeted the sea-floor. Over time, they became cemented together to form the nummulitic limestone.

These particular limestone formations have become exposed in many different places. Where the distinctive coin-like fossils have eroded out of the bedrock in North Africa to be scattered among the desert sands they are known to the Bedouins as 'desert dollars'.[3] And in the Crimean Peninsula, craggy outcrops of this nummulitic limestone form the very jaws of the 'Valley of Death' that witnessed the disastrous Charge of the Light Brigade during the battle of Balaclava in 1854, as memorialised in Alfred, Lord Tennyson's poem.[4]

Thus the enormous blocks of rock that form the Great Pyramid at Giza were quarried from what is essentially a huge slab of limestone stretching across Eurasia and Africa. Composed of countless foram shells, this nummulitic limestone is a biological rock. So while the Egyptian pharaohs may have ordered their construction from enormous blocks of limestone, it was another life form that built the pyramids. The tombs of the pharaohs are made of innumerable drifts of the skeletal remains of a giant single-celled sea creature.[5]

The pyramids are one of the most enduring symbols of human civilisation, revealing what we can build when we put our minds, and coordinated efforts, to it. Throughout history many of the grandest edifices have been constructed out of devotion to the divine: the step pyramids of Mesoamerica, the temple complexes of Sanchi Stupa and Angkor Wat, or the medieval cathedrals across Europe. But the materials with which these monuments were made are the same as those used for buildings constructed for more pragmatic purposes – dwellings, civic buildings, bridges, harbours, fortifications. At the root of all this fervent construction is a fundamental human requirement: to find shelter from the elements. And throughout history we've turned to the natural materials we found around us.

WOOD AND CLAY

Many cultures around the world, especially nomadic peoples, have constructed temporary structures like the wigwam, tipi and yurt out of branches, bark, reeds or animal hides. Timber, of course, is one of the oldest construction materials. A variety of different trees can be worked into supporting beams, poles and planks, as well as cladding slats or roofing tiles. And before metal was widely available, timber was also used for mechanical components.* The cross-grained fibres of elm make it resistant to splitting and so perfect for the hubs of cartwheels. Hickory is particularly hard and so was used for the gear teeth in the drive systems of waterwheels and windmills. And pine and fir trees grow exceptionally tall and straight and so are well suited for ships' masts.

The simplest material suitable for fashioning into solid walls is clay. The early city-dwellers of Mesopotamia, the land between the rivers, lived in a world of mud. Although a perfect environment for productive agriculture, the area is woefully lacking in natural resources like timber, stone and metals, all

---

* Metals, such as bronze and then iron and steel, were initially in such short supply that they were used only for fastening other, more readily available, structural materials together, such as hard nails joining timber beams. It is only with the cheap availability of iron and steel since the Industrial Revolution, and machining techniques for the mass-production of parts, that metal has become a major structural component itself, for example as rebar elements within reinforced concrete or girders supporting bridges and modern high-rise buildings.

of which had to be imported. A succession of ancient Mesopotamian civilisations – Sumerians, Akkadians, Assyrians and Babylonians – survived by trading their food surplus for cedarwood from Lebanon, marble and granite from Persia and Anatolia, and metals from Sinai and Oman.[6] Nonetheless, most of their structures were constructed from what was locally available. Houses and palaces, city walls and forts alike were all built with sun-dried adobe bricks. Even the cores of their great ziggurats – the tiered, flat-topped pyramids that served as temples – were made of sun-baked bricks. More durable kiln-fired bricks were only used for the facing of the palaces and ziggurats and they were decorated with colourful glazing. Mud even became a writing material, when the Sumerians invented writing by pressing a stylus into tablets of soft clay.[7]

In fact, long before providing the ancient Mesopotamians with earthen bricks and soft tablets for the earliest forms of writing, clay had proved transformative to human existence. The innovation of firing clay into earthenware pots gave us brand new capabilities. Pottery provided vessels in which food could be cooked by boiling or frying. Not only does cooking deactivate certain plant poisons that are present in the potato and the cassava, for example, and so make more foodstuffs available to us; it also breaks down complex molecules to release more nutrients that our bodies can absorb. In short, pottery enabled greater processing of food to make it more digestible to us. Lidded containers fashioned from clay also protect food stores from pests and vermin, and make them far more portable for travelling and trade. Pottery can be made more watertight and more attractive by glazing it – coating it in a solution of certain powdered minerals before kiln-firing – and this may well have been what made humanity stumble across the process of smelting metals like lead or copper. We discovered metals because we loved colour.

Fired clay proved critical to our development through history not only because it is hard and watertight, but also because it can be extremely heat-resistant. Firebricks are ideal for lining kilns and furnaces: they insulate the heat inside without being affected themselves, and so allow very high temperatures to be achieved. Thus ceramics enabled humanity to truly master fire, not just for keeping the night-time cold at bay or for cooking, but for taking raw materials gathered from the environment and transforming them into some of the most useful substances of history: smelting metal out of their ores, calcining lime to create mortar, or producing glass.

The Mesopotamians built with dried mud for want of harder, more durable materials. But elsewhere around the world we have used the geology beneath our feet. We not only build our cities *within* the landscape – near the coastline, in a fertile river valley or close to hills with mineral resources – but we also make them *of* the landscape. In this chapter we'll see not just how the Earth made us, but how the planet provided the solid materials that we have used for construction. The story of civilisation is the story of humanity digging up the fabric of the planet beneath our feet and piling it up to build our cities.

There are three fundamental rock types on Earth, and throughout history we've built using all three kinds. Sedimentary rocks are formed by the deposition and then cementation together of material which either eroded from older rocks or was produced biologically – sandstone, limestone and chalk are all examples. Igneous rocks such as granite, on the other hand, solidify from volcanic lava or magma still deep underground. And when sedimentary or igneous rocks are subjected to high temperatures and pressures – caught in the crunch of continental collisions or when magma intrudes up into them – they are transformed physically and chemically, becoming a metamorphic rock like marble or slate.

The ancient Egyptians were the first civilisation to extensively quarry and build with natural stone, and they made use of a variety of different rocks. Nubian sandstone was available from the cliffs flanking the Nile in Upper Egypt. The Great Temple of Rameses II at Abu Simbel and the Luxor Temple in Thebes, for example, were carved of this yellow-brown stone. Further north, the Nile cuts through the nummulitic limestone that overlies the older Nubian sandstone which, as we saw earlier, was quarried to construct the pyramids at Giza. In the Eastern Desert, the rifting open of the Red Sea exposed the ancient basement rocks that form the very foundations of the African continental crust. The granite and gneiss (formed by granite's metamorphosis) here are over half a billion years old. Hard and durable, they were prized by the Egyptians for carving statues and obelisks, and taken by barge down the Nile to be exported around the Mediterranean world.[8]

Let's take a look now at some of the most important rocks we've used through history, and how our planet created them.

LIMESTONE AND MARBLE

As we have seen, the nummulitic stones used to construct the pyramids are a form of limestone. But it is just one kind of this very widespread rock type. Calcium carbonate rock is also created at the mouth of volcanic hot springs where, as the water cools, the minerals precipitate out of solution and rapidly form layers of limestone on the ground. This form of limestone is known as travertine. The main pillars and external walls of the Colosseum in Rome, for example, were made of travertine quarried from Tibur (today's Tivoli, a town about 30 kilometres north-east of Rome), and hot-spring limestone from the same place was used for the Getty Center in Los Angeles.[9]

However, most limestone is not formed in volcanic hotspots on land like the Tivoli mineral springs, but on the seafloor as biological rock. Much of the limestone found across Europe and the rest of the world was formed during the Jurassic Period, as warm, shallow seas flooded the land. Marine reptiles like pliosaurs and ichthyosaurs swam these tropical seas,[10] whilst on the sea floor calcium carbonate from the shells of sea creatures – such as forams – precipitated as a limy mud. As particles of sand or shell fragments rolled back and forth around the seabed with tidal currents they became coated with concentric layers of calcite mineral to form tiny balls called ooliths (from the Greek for 'egg-stone'). These little spherules then became cemented together with more calcite to form oolitic limestone.

In Britain, oolitic limestone formed during the Jurassic has resurfaced as a great sash across the country, stretching from East Yorkshire, through the Cotswolds, and down to the Dorset coast (see map on page 151). Oxford lies in the middle of this band and many of its university colleges were built of this glorious, golden stone.[11] At the very south-western tip of this diagonal stripe of Jurassic limestone lies the Isle of Portland, a promontory jutting out into the English Channel, its hard rock resisting the pounding of the waves. The limestone exposed here dates to the very end of the Jurassic Period, 150 million years ago.

Portland stone is a fabulous building material, and not only for its delightful creamy tint. The ooliths that made it were cemented together in just the right amount: it is durable enough to resist weathering and crumbling, but not impractically hard for masons to cut and carve. Portland stone is known as a freestone: its fine-grained texture can be cut smartly in any direction and it has been used as a building material since Roman times. Portland stone became the rock of choice for many of Britain's monumental or civic buildings. Its pure hue appears in the Tower of London, Exeter Cathedral, the British Museum, the Bank of England, and the facing of the east side of Buckingham Palace (including the famous balcony). Sir Christopher Wren selected it for the rebuilding of St Paul's Cathedral, as well as many other London churches, after the Great Fire of London in 1666. Portland stone has also been used around the world, for example in the United Nations building in New York.

The United States has its own sources of limestone. Some of the highest quality is quarried in southern Indiana where it was laid down, much earlier than Portland stone, during the early Carboniferous Period around 340 million years ago. Indiana limestone has been used in the facing of the Empire State Building, the Yankee Stadium in New York, the National Cathedral in Washington, DC, and the Pentagon. This stone was also used extensively in the reconstruction of Chicago after its Great Fire of 1871, emulating the rebuilding of London's monumental landmarks after its own conflagration two centuries earlier.

Much of the northern Mediterranean coastline that we explored in the last chapter is also made up of limestone rocks originally laid down on the bed of the Tethys Sea. Now thrust up above the waves, it has been dissolved away by rainwater trickling underground, a process which has created an extensive networks of caves. Perhaps unsurprisingly, many of these have become associated with the Underworld of classical mythology. At the tip of the Mani Peninsula, the southernmost point of Greece, for example, is the entrance to a cave where the legendary Orpheus is said to have descended into the Underworld to find his deceased wife Eurydice. The beauty of Orpheus' lyre playing won over the god Hades, who allowed him to take Eurydice back to the land of the living on one condition: that he should not look back. But as soon as Orpheus had reached the upper world he anxiously turned round to check that she was following, and so Eurydice disappeared for ever.[12]

Where this Tethyean limestone has been baked underground in the convergent plate boundaries around the Mediterranean – by magma rising up and intruding into it, or by its being caught in the tectonic vice crunching up mountain ranges like the Alps – it is metamorphosed into marble. This is the signature stone of Classical Greek and Roman sculpture, monuments and grand public buildings. Some of the most highly prized marble in the world is still quarried around the city of Carrara in northern Tuscany. Here the Apuan Alps contain mountains of the pure-white stone which has served as a building material since the days of ancient Rome, when it was used for the Pantheon and Trajan's Column, for example. Carrara marble was also a favourite of Renaissance sculptors: it provided the material for what is probably the most famous statue in the world – Michelangelo's *David*. It has also been exported around the world to construct some of the most iconic global landmarks: Marble Arch in London, the

Peace Monument in Washington, DC, Manila Cathedral, Sheikh Zayed Mosque in Abu Dhabi and Akshardham in Delhi.

And it wasn't just the physical building materials that were exported around the globe. The characteristic architectural components of antiquity – from columns to caryatids, from pediments to pilasters – were emulated in Europe for centuries, from the Renaissance through the baroque to the neoclassical of the mid eighteenth century, and on. It was adopted with particular enthusiasm by the nascent United States of America. After winning independence from Britain, this new nation forged its own governmental system, a federal republic, by drawing on some of the political structures developed by the most powerful republic in Western history, that of ancient Rome. At the same time, the architecture of many major public and municipal buildings in America emulated the styles of antiquity. They were constructed not of limestone and marble originating from the ancient Tethys Sea, but replicated in the same imposing style and purity of hue with rock quarried in the young American nation.*

CHALK AND FLINT

Chalk is another form of limestone, although at first glance its properties couldn't be more different. Chalk deposits can be found on almost every continent[13] and are the distinctive

---

* Thomas Jefferson, the third president of the United States, helped draft the Declaration of Independence, but also worked on the architectural designs of some of the new nation's buildings. For example, he modelled the Virginia State Capitol on the first-century BC Roman temple, the Maison Carrée in Nîmes (which in turn influenced the design of other state capitols across the country), and his design of the library of the University of Virginia with its rotunda and dome emulated the Pantheon in Rome. Neoclassicism is perhaps most prominent in the city founded in 1790 as the new national capital on the banks of the Potomac River – Washington, DC. The Capitol Building (home of Congress), the Herbert C. Hoover Building (headquarters of the US Department of Commerce), the Treasury Building and the DC City Hall are all imposing examples of this neoclassical style. And the White House was designed by an Irish architect based on Leinster House in Dublin – later to become the seat of the Irish parliament – which itself mimicked architectural features of antiquity.

signature of the Cretaceous Period of Earth's geological history. In fact, the very name given to this chapter in our planet's story comes from the Latin word for chalk, *creta*.

A thick layer of chalk underlies much of southern England (see page 151). It shows as outcrops along the backbone of the Isle of Wight, continues east as the hilly ridges of the North and South Downs, and sits beneath London, where it forms a bowl that holds overlying layers of clay. The flat chalklands of Salisbury Plain are home to one of the most impressive monuments of early human habitation of Northern Europe, Stonehenge, begun around 3,000 BC. Although the enormous sarsen blocks making up the main ring consist of sandstone, it seems that the builders were attracted to this area by the flint that could be dug from the chalk landscape for tools like knives and arrowheads. Other, less arduously constructed, but equally eye-catching, monuments have been created in this geological band. Humanity has been exploring the artistic potential of this landscape for millennia, scraping away the thin layer of turf overlaying the porous chalk to reveal the bright white rock beneath, or cutting trenches into the ground and filling them with chalk rubble. Chalk figures crafted on hillsides, visible from miles away, include the stylised outline of the Uffington White Horse in Oxfordshire, created in the Bronze Age,[14] and the proud salute of the Cerne Abbas Giant in Dorset which probably dates to the first century AD.[15]

The chalk layer is most clearly visible on the South Coast, where it forms the eye-catching White Cliffs of Dover. It continues under the English Channel to France, where it has produced mirror-image white cliffs and provided the terroir of the great French wine regions of Champagne, Chablis and Sancerre. The Channel Tunnel carrying high-speed trains between Folkestone and Calais was burrowed 50 kilometres through a chalk marl layer, a muddy chalk deposit that is soft but impermeable. And as we saw in Chapter 2, the chalk bridge that used to physically connect Britain to mainland Europe was scoured away in a cataclysmic flood event.

Some rocks contain beautifully preserved fossils. Along the Jurassic Coast in South West England, for example, where

190-million-year-old mudstones are rapidly worn back by the sea, a pleasant day can be spent strolling along the eroding cliff face hunting for spiral ammonites, bullet-shaped belemnites or brittle star fossils. However, the great chalk layers don't so much contain fossils: rather, they *are* fossils. The White Cliffs of Dover are a 100-metre-high exposed slab of biological rock.

The larger fossils that are visible if you look at a lump of chalk under the microscope, about a millimetre across, are the multi-chambered shells of forams – the same kind of single-celled marine organism that formed the giant nummulitic fossils in the limestone used to build the Great Pyramid. But the bulk of chalk is made up of what looks like a very fine white dust. Zoom in on these powdery particles with a high-powered electron microscope, however, and you'll see that even these have the unmistakable intricate detail of biological shells. These particles come in a variety of forms, but perhaps the most distinctive are the fragments of tiny spheres that look like overlapping ribbed dinner plates. They are the minute armour casings of coccolithophores – tiny single-celled algae found among the floating plankton of sunlit surface waters.

These vast deposits of chalk formed during the late Cretaceous Period, roughly between 100 and 66 million years ago. This was a time of exceedingly high sea levels around the world which were around 300 metres above where they are now. As much as half of the continental land area that is dry today would have been submerged back then. The Tethys Sea rose up to inundate much of Europe and South West Asia, extending its wide arms as great seaways through the centre of North America and into North Africa.

The reasons for these high sea levels weren't just the sweltering conditions of the late Cretaceous that stopped ice caps from forming at the poles – this has been the case for much of Earth's history. They were the result of the frantic activity of the continental break-up at the time. During the late Permian Period, 200 million years earlier when the world's large landmasses had congregated into the huge supercontinent of Pangea, global sea levels were at one of their lowest marks of the past half-billion years. The huge mountain ranges thrown up as the continents crashed and fused together meant that more continental mass was lifted up out of the oceans. But with the subsequent disassembly of Pangea, rifts tore up the supercontinent. First, Pangea ripped apart roughly along its middle as Laurasia moved north away from Gondwana. Later the South and then North Atlantic oceans formed as new spreading rifts tore apart Africa and South America, and North America and Eurasia, respectively. The new, hot oceanic crust formed at these long rifts buoyantly rose up in vast submarine mountain ranges, displacing the surrounding seawater – just like when you lower yourself into a bath. It was this planetary process that caused ocean levels to peak during the late Cretaceous.[16] Warm seas covered wide areas of the continental land, providing boom conditions for the growth of forams and coccolithophores, their tiny shells building up in thick deposits of calcareous sediment on the sea floor; and these became chalk.

Unlike limestone, the soft and crumbly chalk does not generally offer a great building material itself. But it does lend itself to being crushed and spread on agricultural fields to lower the soil acidity, in the production of quicklime for cement and in a whole range of chemical processes. Bricks can be baked out of moulded blocks of clay, but to build a sturdy wall they need to be stuck securely together. It is limestone and chalk that we have learned to use for this construction alchemy. These calcium carbonate rocks are crushed and roasted in a kiln so that they chemically break down (releasing carbon dioxide in the process) before being mixed with water to make a soft putty. In this way, limestone not only provides us with building blocks, but also with the glue for sticking other materials together. Mortar, cement and concrete are essentially artificial rock that can be spread or poured into any desired form and when set becomes hard as stone.

Chalk also contains beds of flint nodules. Unlike the soft, bright-white, almost chemically pure calcium carbonate of chalk, the flints are hard and dark-coloured lumps of silica. While forams and coccolithophores build their casings out of calcium carbonate, other single-celled plankton like diatoms and radiolarians form their hard parts out of silica. When these organisms die, their siliceous carapaces drift to the seafloor and dissolve. This produces a siliceous ooze on the sea bed that then forms into flint nodules within the chalky sediment.

As the soft chalk is weathered, the durable flint nodules are eroded out and remain scattered across the landscape. Flint was incredibly important in Stone Age toolmaking. Like volcanic obsidian that, as we saw in Chapter 1, was used for many of the earliest implements in humanity's cradle within the Rift Valley, flint can be knapped to create a very sharp edge or point, perfect for butchering a kill, skinning and scraping animal hides to prepare into clothing, shaping wood, or creating knives, spear points and arrowheads. And flint has remained important ever since. Glass-making requires high-purity silica, and flint offers one such source. For example, flints from South East England were used by George Ravenscroft in 1674 for his lead crystal glassware.* This lustrous glass was produced to rival that of Venice, where craftsmen obtained their silica by roasting white quartz pebbles picked from the bed of the river Ticino, flowing down from the Swiss Alps.[17]

FIRE AND LIMESTONE

We've explored so far how rocks like limestone and chalk have defined landscapes, and provided the raw materials for construction in the form of masonry blocks and as ingredients for

---

* The term 'crystal glass' is something of a misnomer – the amorphous atomic structure of glass is in many ways opposite to that of the strictly regular repeating pattern of a crystal.

mortar, cement and concrete. We build with these materials to protect us from the elements, but the very creation of this biological rock may also have helped to protect life on Earth from the threat of cataclysmic mass extinctions.

One of the greatest spasms in the history of life on our planet occurred at the boundary between the Permian and Triassic periods, 252 million years ago. This end-Permian global extinction event happened when all the world's landmasses were fused together into the single supercontinent Pangea, and it was by far the worst mass extinction to have occurred in the half-billion years of complex life on Earth. The fossil record reveals that around 70 per cent of all terrestrial and up to 96 per cent of marine species were wiped out in this apocalypse and it took the world's biodiversity almost 10 million years to recover.[18] This global wiping clean of the slate also marked a fundamental shift in the characteristic life forms on Earth: the era of 'old life' (the Palaeozoic) gave way to that of 'middle life' (the Mesozoic) – an age that came to be characterised by dinosaurs and gymnosperm conifer trees.*

The cause of the Permian Great Dying is thought to have been massive outpourings of lava. Several pulses of extensive volcanism disgorged a total volume of perhaps 5 million cubic kilometres of runny lava that flowed for hundreds of kilometres, covering huge areas of land with seas of the hot stuff which then cooled and set as large regions of basalt rock.[19]† As these regions were flooded again and again with lava, layer upon layer of basalt built up. They can be seen today as the extensive mountainous plateaus of the Siberian Traps: the hundreds of layers stacked on top of each other

---

* Similarly, the mass extinction at the close of the Cretaceous Period 65 million years ago ended the Mesozoic and ushered in the era of 'new life' (the Cenozoic). It created our world dominated by mammals and angiosperm flowering plants that we encountered in Chapter 3.

† By comparison, the largest eruption in the last millennium, that of the Tambora volcano in 1815, released only 30 cubic kilometres of material – 160,000 times less.[20]

resemble a staircase and so were named after the Dutch word for stairs, 'trap'.*

Such extensive volcanic eruptions would have released huge amounts of carbon dioxide into the atmosphere. Moreover, geologists reckon that the magma gushing out to form the Siberian Traps may have been supercharged with volcanic gases by two other factors. It is thought that as the mantle plume rose up from deep in the Earth's interior beneath Siberia it melted some ancient oceanic crust that had been previously swallowed by subduction. This recycled crust was rich in volatile compounds and so released a large amount of gas when heated. It also appears that on their way up to the surface through the overlying crust these flood basalts encountered strata like coal seams, which the magma baked to high temperatures to release yet more gas.

It seems likely, therefore, that the onset of the outpouring of the Siberian Traps wasn't like any volcanic eruption we'd be familiar with today, but began with colossal belches of gases from the belly of the Earth. The huge volumes of carbon dioxide released by these eruptions created a powerful greenhouse effect. The Earth's surface temperature rose rapidly, and the deeper ocean waters became anoxic – lacking in oxygen – asphyxiating life on the sea floor. Other noxious volcanic gases, such as hydrogen chloride and sulphur dioxide, may also have been projected high into the stratosphere. The output of hydrogen chloride would have severely depleted the ozone layer, allowing harmful ultraviolet rays from the sun to reach our planet's surface. And the sulphur dioxide would have acted to partially block the sunlight, hampering photosynthetic life and the other life forms supported by it, before precipitating back out of the atmosphere again as acid rain.

It's this multi-whammy at the end of the Permian that rapidly collapsed ecosystems across our planet and triggered the largest mass extinction in the history of complex life on Earth. And

---

* This sense also survives in the English language; originally a trapdoor was one that opens to stairs.

the phenomenon wasn't limited to the Permian: another flood basalt event around 200 million years ago, at the juncture between the Triassic and Jurassic periods, is believed to have caused the mass extinction that cleared the way for the dinosaurs to become the dominant land animals.

But then something curious happened. There have been a number of other large flood basalt eruptions since the Permian and Triassic events, yet none of them seems to have triggered a similar mass extinction. Something must have changed on our planet to have made the Earth much more resilient to the potentially cataclysmic effects of mega-eruptions.*

Two huge outpourings of lava about 60 million and 55 million years ago created the North Atlantic Igneous Province as North America rifted away from Eurasia, constituting the final cut in the break-up of Pangea. The basaltic rocks from this event – the distinctively geometric columns of Giant's Causeway in Northern Ireland and corresponding features in eastern Greenland – have become separated by the opening of the North Atlantic Ocean. These outpourings of lava probably released even more molten rock than the Siberian Traps during the Permian extinction.[21] And like the Permian flood basalt eruptions, the magma spewing out to form the North Atlantic Igneous Province also passed through volatile sedimentary rocks near the surface which would have released vast amounts of carbon dioxide as they were baked, in addition to that given off by the volcanic lava itself.[22]

But these events triggered no mass extinction. There certainly was a shock to the Earth's climate, and the second phase 55 million years ago coincided with the Palaeocene–Eocene Thermal Maximum we looked at in Chapter 3. Yet although a few

---

* The end-Cretaceous mass extinction – that saw the death of the dinosaurs, along with three-quarters of all marine species – coincides with the eruption of the Deccan Traps onto India. This occurred 66 million years ago as the subcontinent was gliding north to its eventual collision into Eurasia, passing over a rising magma plume as it burst to the surface. The final straw for life was the impact of a 10-kilometre-wide asteroid or comet slamming into the Gulf of Mexico at the same time.

deep-sea species did perish during this temperature spike, these events seem instead to have stimulated the rapid evolution of the three major orders of mammals that dominate the land today: the artiodactyls, perissodactyls and primates.

So what is it about the Earth since the Jurassic that has made our planet so much more resilient against mass extinctions from large flood basalt events?

One important factor is – again – the break-up of Pangea. Supercontinents are on the whole less effective at removing carbon dioxide from the air. Large areas of interior land far away from the sea become very dry with low rates of rainfall. This means less CO_2 is being scrubbed by the erosion of rocks, and fewer rivers carry sediment and nutrients into the ocean to fertilise plankton growth, thus also suppressing the biological mechanism for absorbing CO_2. So in the last 60 million years, since the final break-up of Pangea, the world has been more effective at removing carbon dioxide released into the atmosphere by large outpourings of lava. But this can't be the whole story. The geological mechanism for lowering carbon dioxide in the atmosphere – by the erosion of mountains – works very slowly. Thus the sudden leap in CO_2 resulting from the eruption of a large igneous province would trigger a mass extinction long before rock erosion was able to bring the levels back down again. It seems that the important factor was a crucial biological transition.

During the early Cretaceous Period, around 130 million years ago, coccolithophores expanded out from the shallower waters of the continental shelves to live as plankton in the open ocean. Around the same time calcite-shelled forams also spread from their deep seafloor habitat to the surface waters of the seas. This meant that the vast open ocean itself, and not just the shallower waters around the continents, hosted plankton that produced calcite shells. When shells from dead coccolithophore and foram plankton rained down onto the seafloor they formed a new kind of sediment, creating limestone in the ocean deeps and not just on the continental shelves.[23] Thus marine life was becoming much more adept at removing carbon dioxide from the atmosphere and locking it away in biological rocks on the deep sea floor. And since this time, the carbon dioxide levels on our planet have been steadily diminishing.

Now even with the sudden injection of huge amounts of carbon dioxide into the air from flood basalt events, the oceans' limestone-forming plankton were able to scrub this gas out much more rapidly than any geological processes. Since the early Cretaceous, therefore, the Earth has developed a powerful compensation mechanism for rapidly removing sharp rises in volcanic carbon dioxide before it can trigger runaway warming and mass extinctions. So when 55 million years ago the Palaeocene–Eocene Thermal Maximum started pushing carbon dioxide levels and global temperatures towards catastrophe, plankton saved life on Earth.

Thus the biological rock of the White Cliffs of Dover and the limestone facade of the United Nations building can both serve as reminders of the deep connections within the Earth that across time have created the world we inhabit today.

TECTONIC SWEAT

Granite is the most common rock type of the continents. As we have seen, oceanic crust is formed of basaltic rocks that have solidified from fresh magma seeping out of rifts spreading on the sea floor. But granite is instead forged at convergent boundaries where tectonic plates are forced together.

As oceanic crust is subducted, the water-bearing rocks of this descending plate are melted by the considerable pressure and temperature at depths of between 50 and 100 kilometres, while also being heated by the grinding friction as they slide underneath. This molten magma rises up into the overlying crust and pools into huge subterranean chambers. Here it begins to cool, and as the first minerals crystallise and sink out of the mixture – those with the highest melting point – the chemical composition of the melt left behind in this deep cauldron slowly changes. The early-forming minerals are low in silica (silicon dioxide), which means that the remaining magma becomes more and more enriched in it. Granitic magma is also formed when continents collide and the crust is thickened beneath the great mountain range that is created, partially melting at the bottom and again rising up through the overlying crust. When this silica-rich magma cools and solidifies it forms great subterranean masses of granite rock, often within the core of the mountain range formed above it by the same convergent tectonics. Granite is the sweat of plate tectonics.[24]

This re-melting and chemical processing of the crust also means that granite is less dense than basalt. So in the recurring clashes of plate tectonics the granitic rocks ride over the heavier oceanic basalt and don't become subducted – they survive and accrete together as the basement layer of the continental crust. Thus granite forms the very foundations of the continents, lying beneath the veneer of sedimentary deposits, and only becoming exposed on the surface as austere outcrops when the softer landscape has been eroded around it.

As we have seen throughout this book, no sooner have they been thrust up into the skies than mountain chains experience the punishing forces of the planet that work to rub them away again. The expansion and cracking of freeze–thaw cycles cleaves and pulverises their rocks; rivers coursing down their flanks gouge out great valley networks; and driving glaciers scour the very summit, picking up and rasping fragments of the mountain's own substance to further grind it down. But as mountains are eroded away, the weight pushing their thick crustal roots into the dense mantle is reduced and so they buoy back upwards a little more. The diminishing peaks are therefore relentlessly raised back into the grinding maw of erosion, like a block of wood that a carpenter smoothly pushes into a spinning sanding disk to wear it down. In the end, even the mightiest mountain chain is disassembled grain by grain over the expansive gulf of time of our planet's history. Eventually, the mountains will be worn down to the merest stump, exposing their heart of hard granite.

So when you stand on a pillar of granite, you are stepping on the very core of an ancient mountain range. During its formation, this granite would have had at least 10 kilometres of rock piled on top of it, now worn away over 100 million years or more of erosion. The tors of Dartmoor, El Capitan in Yosemite National Park, Rio de Janeiro's Sugarloaf Mountain and the Towers of Paine in Chile were all created and then revealed in this way.[25]

Granite is hard and durable, with a coarse-grained texture from the large crystals that had time to grow and develop as the melt cooled slowly deep underground. Since granite epitomises solidity and permanence, we have used it to build impressive monuments throughout history. Perhaps the most famous granite feature in the world is Mount Rushmore, in South Dakota. This granitic mass formed 1.6 billion years ago, and in the 1930s the faces of four US presidents – Washington, Jefferson, (Theodore) Roosevelt and Lincoln – were sculpted into its south-east side, so as to catch most sunlight. (The project as originally conceived was to carve the presidents' forms down to the waist, but funding ran out.) The granite of this sculpture is exceedingly hard-wearing, and erodes at the rate of only about 2.5 mm a millennium – it will remain a symbol of American ideals for a very long time. In fact, the designer of the monument took this into account and had the presidents' features carved several inches thicker, so that they will have worn back to their intended shapes 30,000 years in the future.[26]

In the ancient world, the Egyptians were the masters of granite working, sourcing the material from the upriver Nile Valley in quarries in Nubia, in what is today northern Sudan.[27] This they carved into their most enduring columns, sarcophagi and obelisks, such as the 'Cleopatra's Needles' that now stand in London, Paris and New York (although a misnomer, as they were made over 1,000 years before the rule of Cleopatra).*

---

* Indeed, Cleopatra lived closer in time to the modern world of iPhones and the glass pyramid of the Louvre gallery in Paris than she did to the ancient construction of the Great Pyramid at Giza.

It was the rediscovery of the ancient Egyptian monuments and their display in the British Museum that inspired European stonemasons in the early 1800s to attempt to emulate their works and carve granite, only becoming successful with the development of steam-powered machinery for cutting and dressing granite in Aberdeen.[28] Much of the granite used in Britain is sourced from Aberdeen, where it had been formed beneath the great Grampian mountain range 470 million years ago[29] – long enough ago for erosion to rub away the kilometres of overlying rock to reveal the granitic core.

Even the durable resilience of granite is not impervious to the merciless action of the elements, though. As it reacts slowly with water, granite is chemically rotted and undergoes an almost magical transition. The quartz crystals tumble away as grains of sand, and another mineral component of the original granite, called feldspar, is chemically converted into kaolin, a type of clay. The water leaches out other impurities from the decomposing granite to leave behind only the fine, flakey particles of this purest of clays, snow-white in appearance. This can happen when the deep granite has been slowly exhumed and exposed to the elements, or while it is still underground and its own heat drives hydrothermal systems in the subterranean cracks and fissures.*

* This is the process by which almost all sand in the world's beaches and deserts was formed.[30] Quartz is also the base material we use today for making glass, and we refine it into the extremely pure silicon wafers of microchips and solar panels. Quartz didn't exist on the early Earth – it was created by the action of plate tectonics over hundreds of millions of years. We saw earlier that convergent boundaries cause the crust to melt and form vast magma chambers. As the magma cools in these huge cauldrons, the first minerals to form leave the remaining magma with a higher and higher proportion of silica, which then crystallises as granite. While the primordial composition of the deep mantle is 46 per cent silica, the granite produced by this magma differentiation process has been enriched to about 72 per cent silica, high enough for crystals of quartz (pure silica) to form. Thus plate tectonics on our planet are like a chemical processing plant, acting to purify silica over time, and so make it available for human technologies. Incidentally, this means that if the Earth-like planets we are now discovering orbiting other stars don't have plate tectonics, they may well have warm oceans, but no sandy beaches.

Not only is the kaolin a pure, snow-white colour, but its powdery, plate-like particles make it wonderfully soft and malleable. This clay can be fired at high temperatures, creating pottery that is particularly strong and also translucent. Kaolin is therefore the raw material for the very finest of ceramics – porcelain.

Porcelain was first developed by the Chinese around 1,500 years ago, and reached the Islamic world in the ninth century AD. The trade of porcelain into Europe gave it its name in English: fine china. The firing of porcelain vases, jugs, bowls and tea sets at high temperature makes them strong even when very thin to give them a refined delicacy and an almost ethereal translucency. It's this that made porcelain so highly prized in comparison to other clay ceramics – earthenware or stoneware retain their opaque muddy colour even when colourfully glazed.

When trying to emulate porcelain, English potters added the ground ash of bones from their abattoirs, but although reproducing the white colour this bone china was still inferior to porcelain. They eventually discovered the secret ingredient of kaolin clay, and the first commercially successful production in England was achieved in Stoke-on-Trent in the final years of the eighteenth century. The area has abundant coal for firing pottery kilns, and the Staffordshire potteries originally made use of the clay deposits found between the local coal seams, firing them into building bricks, floor tiles or huge pots for transporting butter down to London by packhorse.[31] But with the development of techniques for manufacturing fine bone china, Stoke-on-Trent became the leading production centre in Europe for this rival to porcelain. Yet although the Stoke potteries had abundant coal for their kilns nearby, and came to use it for steam engines for crushing and mixing the raw materials and driving the potters' wheels, they needed to import the crucial kaolin from Cornwall. Like Aberdeen, Cornwall has exposed granite formations, but here the rock has been hydrothermally processed into the soft white kaolin clay. And the demand for sending Cornish kaolin to the potteries of Stoke, as well as for transporting the delicate, finished china around Britain, was one of the major drivers behind the digging of the network of long canals in the early stages of the Industrial Revolution.[32]*

In this way granite, formed as the slow-cooling sweat of the crushing pressures and heat of plate tectonics, both gives monuments their enduring solidity and is converted into one of the most delicate and fragile of substances – porcelain china.

THE GROUND BENEATH OUR FEET

We saw earlier in the chapter how the ancient Egyptians and Mesopotamians constructed their civilisations with the local endowment of building materials provided by the underlying Earth. This is as true through modern history as it was for the earliest civilisations. Let's explore how the normally unseen, subterranean world is reflected in the appearance of buildings across Britain, the place where the first nationwide geological map was drawn.†

The geology of Britain is particularly diverse, displaying outcrops of rock from almost all ages of Earth's history over the past three billion years. Tectonic shifts and erosion have, over time, re-exposed these different strata in complex swirling stripes across the country. In age they trend roughly north to south, from the most ancient rocks in the Scottish highlands to

---

* Today, one of the old kaolin quarries in Cornwall holds the Eden Project. This ecological tourist centre was constructed as a cluster of geodesic domes made of inflated plastic bubbles. These are innovative greenhouses that host tropical and Mediterranean biomes, and huddled within the crater-like pit they look almost like a sci-fi colony on Mars.

† Whilst inspecting excavations for coal mines and canals in Somerset, the surveyor William Smith realised that different layers of rock were always found in the same sequence underground, and that these strata could be identified by the fossils they contained. He travelled across Britain to survey the strata exposed by natural escarpments and the quarries, canals and railway cuttings of the Industrial Revolution, resulting in 1815 in his geological map of Britain that shows the different rock strata present near the surface.[33]

the youngest formations, created over the last 65 million years, in the South East. It is fascinating to see how throughout history the characteristics of buildings across Britain generally reflected the local geology: we recognise the dark granite of city buildings in Aberdeen and of farmhouses around Dartmoor, the buff-coloured Carboniferous sandstone of Edinburgh and

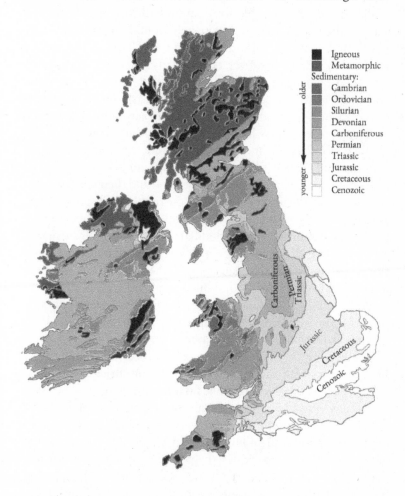

Geological map of the British Isles.

Yorkshire, the golden Jurassic limestones of the Cotswolds villages, and the warm brown colour of clay used for bricks and roofing tiles in and around London. We have pulled the geology from beneath our feet and piled it into walls, and just by looking at a photo of a traditional building a geologist would have a good idea in which part of Britain it was taken.

Places with no suitable local stone had to make do as best they could. Chalk does not make a great building material – it's a soft and crumbly rock, and doesn't fare well in the face of weathering. It has, however, occasionally been used as a material called clunch, in the form of irregular lumps of rubble or cut into blocks and laid in courses – in East Anglia as well as in Normandy, for example. But generally across the Cretaceous landscape alternatives had to be found. Many cottages in the chalklands of Suffolk and Norfolk were constructed from timber frames, which were filled in with wattle and daub – a lattice of twigs covered with wet soil and straw – and then whitewashed with a solution made with chalk. These timber frames are sturdy, and if properly protected from damp, are durable enough to survive for centuries. As the chalklands also offered little material for roofing tiles, buildings in this geological area were traditionally thatched with reeds or else the long straw left over after the wheat harvest. So, while such timber-framed, thatch-roofed buildings have come to epitomise the quintessentially English countryside, it reflects in reality the paucity of the local geology for suitable building stone.[34]

These idiosyncratic building styles became much more homogenised with the Industrial Revolution. Bricks were mass produced for constructing mills, factories and workers' housing in the growing cities, and transported much greater distances along the canals and then railways. Slate, which has long been mined from the half-billion-year-old Cambrian rocks around Snowdonia in North Wales, began to be used as a roofing material across the country. Slate is a fine-grained rock that began as sea-floor mudstone before it was squeezed and metamorphosed in the vice of plate tectonics. This forced all its particles to lie along a particular plane, so that a skilful tap with a chisel can split it into thin, perfectly flat slices: it is therefore ideal for making roofing tiles. Welsh slate supplied the expanding industrial cities throughout the nineteenth century and to this day these thin wafers of the Cambrian Period cap buildings across Britain.[35]

The rocks of different regions around the world have been important not just for providing the raw materials for our construction projects throughout history: the underlying geology has also determined how our modern cities have developed.

If you can remember a trip to Manhattan, or visit it now with Google Earth, you'll see that there are two main areas of towering skyscrapers: the dense cluster of the downtown financial district on the southern tip of the island; and Midtown, sporting the Chrysler Building, Empire State Building and the Rockefeller Center. Between these two nodes of ultra-highrise edifices lies a spread of lower buildings. It was first argued by a geologist in the late 1960s that this distribution of buildings echoes the invisible strata beneath the streets.[36]

Lumps of a dark, hard metamorphic rock known as schist – originally mud or clay transformed in the crushing heat deep in Earth's interior – outcrop all over the city; New Yorkers on their lunch break might sit on a slab of it in Central Park while they munch their sandwiches. New York's schist was baked beneath a huge chain of mountains running along the eastern United States from the Labrador coast down to Texas, as well as to eastern Mexico and Scotland (before the North Atlantic Ocean rifted open). This Grenville mountain range ran down the middle of a supercontinent even more ancient than Pangea, called Rodinia. Around 1 billion years ago the continent of Laurentia collided with two other continental plates, fusing them together and crumpling up the Grenvilles. Over aeons of time since, whilst the continents have split and recombined in different configurations, the slow but persistent erosion has whittled down this mountain range so that only its base remains today.

In New York the schist rocks exist in a syncline, a trough-like dip underground that brings the schist layer close to the surface at the southern tip of Manhattan, and again in Midtown. This hard metamorphic bedrock provides the perfect foundation for bearing the immense weight of soaring skyscrapers. In between, held in the cup of the schist syncline, lies softer rock less supportive of massive buildings. Socioeconomic factors also played a role in the pattern of skyscrapers, with development occurring in already established commercial centres, for example, but on the whole the skyline of Manhattan follows the under-lying geology: the areas with the tallest buildings are supported by the hard schist. The invisible subterranean world – the worn-down stumps of a truly ancient mountain range – is reflected above ground in the towering skyscrapers in the commercial districts: monuments not to the gods but to capitalism.[37]

London is in some respects the opposite of Manhattan. Rather than an island bounded by two rivers, it is a city built around a river. But it is located in a similar geological setting. The wedge-shaped London Basin sits at the bottom of a syncline, where the rock layers have been buckled into a trough – in this case by the tectonic forces that also crumpled up the Alps. Indeed, the London Basin is part of the same rippling of surface rocks as the bulge of the Weald–Artois anticline that once formed the land bridge between Dover and Calais that we explored in Chapter 2. While the syncline in Manhattan brings hard, metamorphic schist rocks close to the surface in Downtown and Midtown, London and the whole of the lower Thames Valley runs along the floor of the syncline trough. This became filled with a layer of clay when a warm shallow sea lapped into the tapering depression around 55 million years ago.

This London clay is decidedly unaccommodating for constructing the tallest buildings of the modern age. The reason why London, in contrast to New York, has so few skyscrapers is this thick layer of soft, putty-like clay beneath the city. Towers like The Shard or One Canada Square in Canary Wharf had to be built with very deep-piled foundations to support their weight. The thick clay layer is, however, ideal for digging tunnels: it is soft to bore through, but forms a stable and water-impermeable blanket for the tunnel.

London built the world's first underground metro line in 1863, and today, the Tube has been developed into a network of over 400 kilometres of lines, serving 270 stations (although not all underground). The underlying geography also explains why North London is so well served by the Tube network, but the south has far fewer lines. South of the Thames the clay layer dips to beneath the depth of the network, and tunnels must instead be bored through much trickier strata of sand and gravel. The London clay is also the reason why the Tube has become so uncomfortably hot. Underground caves are normally refreshingly cool, so this would appear as something of a paradox. In fact, when the tunnels were first dug the temperature of the clay was around 14 °C – indeed, in the early days the Tube was advertised as a place to keep cool on a hot summer's day. But after more than a century the heat released by the trains' motors and brakes – as well as by the millions of passengers – has been absorbed into the tunnel walls. And as the dense clay is a remarkably good thermal insulator this heat has found nowhere else to go.[38]

So while the first true cities in the world, in the muddy plains of Mesopotamia, were constructed of sun-dried adobe bricks, underlying clay continues to direct how our modern metropolises develop: the extensive underground Tube network of London contrasting with the towering skyscrapers of New York.

Let's turn now from how the geology beneath our feet has offered us the natural fabric for building our civilisations and cities, to how humanity learned to extract from the rocks the materials for the tools and technology with which we transformed our world.

Chapter 6

Our Metallic World

We've seen how humanity's earliest tools were constructed from stone – by knapping lumps of chert, obsidian or flint – or made of wood, bone, leather and plant fibres. As we progressed through the Palaeolithic, Mesolithic and Neolithic periods (the Old, Middle, and New Stone Ages) we refined these technologies, moving from making chunky hand cutters and scrapers to small sharpened stone flakes suitable for spearpoints and arrowheads. Yet the beginning of the Bronze Age marked a deep transition in the human story: rather than making tools by simply reshaping what could be picked up from the natural world around us, we learned how to purposefully *transform* raw materials, extracting shining metals out of their rocky ores, forging and casting them, and perfecting alloy mixtures. And the rate of technological innovation accelerated over time. It took 3 million years from hominins making chipped stone tools for humans to smelt the first copper; yet we progressed from the Iron Age to space flight in just 3,000.

Metals have been so revolutionary in human history because they offer a range of properties that no other materials provide. They can be extremely hard and strong, but unlike brittle ceramics or glass they are also flexible and shatter-resistant. For our more recent technologies, they are able to conduct electricity and resist the searing temperatures to which high-performance machinery is exposed. And over the past few decades we've come to exploit a staggering diversity of metals for our latest technology, especially in modern electronic devices.

In this chapter, we'll explore how metals have transformed human society from the Bronze Age to the Internet Age, and how the Earth provided them for us.

ENTER THE BRONZE AGE

The first metal we smelted for crafting tools and weapons was copper. Copper ores are often easy to spot – containing minerals coloured an attractive blue or green – and the metal is easy to smelt: it can be extracted by roasting lumps of the ore with charcoal in the same sort of kiln that is used for firing pottery. The burning charcoal provides both the high temperature needed, and also the 'reducing' chemistry to strip the metal from the oxide, sulphide or carbonate it is bonded with in the rock to leave pure copper.

The problem with pure copper is that it is a pretty soft metal: the edges of tools hammered from it dull easily and must be constantly resharpened. A far superior material is offered by mixing copper with another metal to make an alloy – bronze. When larger atoms are interspersed among those of copper they stop the metal being as pliable: they essentially jam the layers of copper atoms from slipping past each other so easily, making the metal mixture harder and more durable. The earliest bronze to be made was an alloy of copper and arsenic, but this mixture was improved on with the copper-tin bronze that was first produced in Anatolia and Mesopotamia in the late fourth millennium BC and then spread to Egypt, China and the Indus Valley.[1] One particular advantage of copper-tin bronze is that it melts at a much lower temperature and does not bubble, and so can easily be poured into a mould for casting.[2] This enables craftsmen to form any shape of implement required, and then repair or even recast it when it becomes worn or gets broken.[3] Bronze soon became the standard material for making ceremonial objects, cooking utensils, agricultural tools and weapons.[4] The Neolithic had given way to the Bronze Age.

The pioneering use of bronze in Mesopotamia is something of a surprise as this region does not have its own sources of tin, so this crucial ingredient of the alloy would have been traded over long distances. The tin used in western Eurasia during the Bronze Age came from mines in the Erzgebirge Mountains along the modern German–Czech border,[5] from Cornwall and, to a small extent, from Brittany. The mines in Cornwall in particular came to offer the ancient world a large supply of the tin they needed. Ores of this vital metal were created when granite magma intruded up into sedimentary rock layers. The heat of this bulk of magma drove hydrothermal systems underground, with the circulating hot water dissolving metals from the surrounding volume and then redepositing them in fissures and fractures in the overlying rock as veins of rich ore.[6]

We know that tin was traded by sea from Northern Europe by the Phoenicians sailing through the Strait of Gibraltar from about 450 BC, and before this along overland trade routes to the Fertile Crescent.[7] And as tin was scarce in the ancient world, it would have commanded a high price. Copper ore, on the other hand, was more widely distributed, and our planet has made it available to us through a particularly intriguing process.

FROM SEA FLOOR TO MOUNTAIN TOP

Bronze Age craftsmen in the Mediterranean, Egypt and Mesopotamia drew heavily on copper mined from Cyprus.[8] Indeed, the island gave its name to the Latin word for copper – *cuprum* – and hence its modern element symbol, Cu. We saw in Chapter 4 how the geological set-up of the Mediterranean created the perfect milieu for thriving seafaring societies, and the tectonic processes in this corner of the world also provided the crucial raw materials for building civilisations during the Bronze Age.

Copper, along with other metals like zinc, lead, gold and silver, is deposited in high concentrations at mid-ocean spreading ridges,[9] where tectonic plates are being heaved apart and magma wells up to form new oceanic crust. Right along the length of these opening cracks in the shell of the Earth hot magma oozes very close to the surface. Seawater trickles and soaks down through the rocks of the sea floor where it meets this magma and becomes superheated. It then surges back up through the crust, leaching minerals out of the surrounding rock as it goes, before it gets squirted forcibly back out of the seabed in hydrothermal vents. When this mineral-rich hot fluid hits the frigid ocean water, particles of metallic sulphide minerals precipitate in a churning, thick, inky dark plume, giving such hydrothermal vents their far more evocative nickname: black smokers. These black smokers form clusters of tall chimney-like structures, like some Gaudí-inspired industrial landscape, in the pitch-black ocean depths.

Black smokers serve as oases on the barren deep sea floor for some of the most extreme life forms on Earth. These exotic communities living far beyond the reaches of sunlight include giant two-metre-long tube worms that were utterly new to scientists when the first black smoker fields were discovered by submersible in the late 1970s, as well as pale-white shrimp, snails and crabs. These sunless ecosystems are powered by microbes able to grow on inorganic energy sources like the metals and sulphides spewing out of the vents.

The particles spurting into the ocean settle back down to smother the area around the vents in high concentrations of valuable metals – copper, cobalt, gold and others – deep on the sea floor, but they are currently inaccessible to miners. It takes special circumstances to make these metal deposits available to us.

As we have seen, at convergent plate boundaries, where two tectonic plates butt head to head against each other, thick layers of sea-floor sediments become folded into mountain ranges. As a result, fossils of sea creatures are commonly found in the mountain peaks of the Himalayas or Alps, for example. These were once ascribed to mythological events like Noah's great flood, until we came to understand the awesome earth-moving power of plate tectonics. But the oceanic crust itself, holding ancient black smokers, is composed of dense, basaltic rock and is almost always subducted beneath the lighter continental crust and swallowed into the depths of the Earth. Every now and then, however, rare slivers of oceanic crust escape being dragged down, and are instead squeezed up and pushed over the continental crust.[10] This appears to happen more frequently with the smallest plates, just as in the Mediterranean with its plate fragments that got caught between Africa and Eurasia as the two crunched together. And this is exactly what has happened on the island of Cyprus.

The oval-shaped mound of the Troodos mountains in the centre of Cyprus is the best example in the world of an ophiolite – a slice of oceanic crust that became beached on top of the continental.[11] This oceanic crust was created in deep water about 90 million years ago, at a spreading rift in the Tethys Sea,[12] and became scooped up on top of Cyprus with the closure of the Tethys as Africa pushed up into Eurasia. The Troodos mountains haven't been significantly deformed and so this ophiolite shows a beautifully preserved cross-section of the layers within the oceanic crust;[13] there are even recognisable fossils of tube worms and snails alongside the ancient hydrothermal vents. Troodos is like a gently mounded layer cake, and as the mountains eroded down, these layers have been exposed in concentric rings. The highest peak in the centre is made up of mantle rocks that would normally be found over 10 kilometres beneath the sea floor.[14]

The Troodos ophiolite offers geologists a perfect opportunity for studying how new ocean crust is formed at a spreading rift (which is of course tricky to watch in action at current constructive plate boundaries like the Mid-Atlantic Ridge). But for Bronze Age civilisations it also made conveniently accessible the metals that had been spewed out by ancient deep-sea black smokers. With a chunk of oceanic crust dumped onto land, miners on Cyprus were able to dig into the relevant level on the mountainside where the metal deposits were located. Indeed, Troodos offered fabulously enriched ores, containing up to 20 per cent copper.[15]

From the second millennium BC Cyprus became the major supplier of copper to Mesopotamia, Egypt and the Mediterranean world.[16] As we have seen, in the Bronze Age charcoal was used to bake the metal out of its ore in copper smelting furnaces, and so Cyprus's productivity also depended on a large supply of timber. In fact, by studying the 4 million tonnes of slag heaps on the island that were dumped as waste once the copper metal had been extracted from the mined rock, archaeologists have been able to calculate how much timber was required. It turns out that throughout the 3,000 years of copper production on Cyprus, the entire area of pine forests covering the plains and the mountainsides of the island would need to have been cleared at least sixteen times over[17] – an early example of sustainable woodland management.[18]

Much of the Cypriot copper was traded by the Minoans, the first major civilisation in Europe.[19] Based on the island of Crete, but with trading posts across the eastern Mediterranean, the Minoans thrived for over a millennium from about 2700 BC.[20] We don't know what these people actually called themselves – the term Minoan was coined by archaeologists in the early twentieth century after the Greek myth of King Minos (with his labyrinth and minotaur) who was believed to have lived on Crete.[21] The Minoans built large multi-storey palace complexes and were also expert at water storage and distribution, enjoying well-developed wells, cisterns and aqueducts long before the Romans – and the world's first flushing toilet at the royal palace in Knossos.[22] But above all, they were master bronze-workers and mariners, spreading their cultural influence through their prowess at sea and trade networks that stretched across the eastern Mediterranean.[23] Most of the bronze artefacts and tools that the Minoans produced and exported were made with copper mined on the nearby island of Cyprus. The Minoan civilisation grew rich by trading this metal wealth and shipping it around the known world. But as we have seen in the case of Iran, enjoying the spoils of plate tectonics can have a nasty flip side.

The same subduction boundary that created the rich copper deposits on Cyprus runs past Crete, forming a deep trench lying just 25 kilometres south of its shores. One consequence of subduction is that the plunging plate releases blobs of molten rock that rise back to the surface to feed an arc of volcanoes. This row of volcanoes forms directly above the melting point in the mantle and so appears on the surface at a characteristic distance downstream from the subduction line. The Hellenic arc is located about 115 kilometres north of the Cretan trench, and here the volcanic peak of Thera – known today as Santorini – poked above the lapping waves of the Aegean Sea. This active volcano has been erupting sporadically for thousands of years, but at some time between 1600 and 1500 BC Thera abruptly detonated in one of the most violent eruptions in history.

The eruption almost completely destroyed Thera – the submerged caldera left behind is a mere husk of the original mountain – and the vast plume of pulverised rock hurled up into the sky covered Crete entirely with ash. Ports like Amnissos on the northern coast, facing Thera across 100 kilometres of open sea, were devastated, buried by volcanic pumice rock swept there by a tsunami triggered by the explosion. But much like the eruption of Vesuvius destroying the Roman cities of Pompeii and Herculaneum one and a half millennia later, for archaeologists the catastrophe served to capture a snapshot of Minoan life at the time, preserving their distinctive writing, ceramics, artworks and architecture.*

---

* As we explored in Chapter 4, much of the northern Mediterranean is volcanic, driven by the subduction of the African tectonic plate beneath the Eurasian. Despite the hazards of volcanic eruptions, however, volcanism also offers opportunities. Not only is volcanic soil rich and fertile for agriculture, but the Romans discovered that the properties of volcanic ash lent themselves to the manufacture of 'pozzolanic' cement. This was used for constructing everything from sea harbours (it set even when poured underwater) to aqueducts and the huge dome of the Pantheon, and the durability and mechanical strength of this Roman cement and concrete is still admired by structural engineers today.

It seems that this catastrophic explosion didn't coincide exactly with the collapse of the flourishing Minoan civilisation, although the precise dating of either event is difficult.* But what is clear is that within a few generations of the Theran eruption, Minoan society was in terminal decline: its palaces were destroyed[25] and the island succumbed to invasion by the Mycenaean Greeks.[26] What had made the Minoans so successful was their maritime proficiency and trade and so the sudden loss of much of their fleet and ports to the tsunami following the eruption, as well as the destruction of their major trading port of Akrotiri on Thera itself, would have hit their economic infrastructure hard. It's likely too that the Minoans suffered acute food shortages and even famine with the loss of their fishing boats and the flooding of their agricultural fields by seawater.[27] The natural disaster shifted the balance of power in the region and left Crete vulnerable to Mycenaean conquest. But it was the Phoenicians, inhabiting the strip of land that is now Syria, Lebanon and Israel, who came to dominate the shipping in the Mediterranean (see page 101).[28]

The Troodos Mountains on Cyprus from which the Minoans had sourced their copper form a large, accessible and exceptionally well-preserved ophiolite, but they are not unique. As plate collision closed up the Tethys Sea to create the Mediterranean, other slithers of ancient oceanic crust were squeezed out on top. Ophiolite metal deposits can also be found in bands around the rim of the Mediterranean, within the Alps, Carpathians, Atlas and Taurus mountain ranges. And around the world other ocean-closing events have also thrown up oceanic crust. Some of the largest mines today, such as Rio Tinto in Spain, Noranda in Canada and those along the Ural Mountains in Russia, dig into rich black-smoker metal deposits of copper, zinc, lead, silver and iron.[29]

---

* No written records of the eruption itself have been recovered, although the Minoan Linear A script has never been decoded and this could possibly contain eye-witness accounts.[24]

Copper-tin bronze provided humanity with metal tools, utensils and weapons for around two millennia, before it was superseded by a far superior metal – iron.

FROM WROUGHT IRON TO STEEL

In fact, we have been using iron for tens of thousands of years, not for its metallic properties but as a colourful pigment to adorn and express ourselves. Ochre can vary in colour from brown through yellow to vibrant red, depending on the exact iron oxide mineral and how much water it contains in its structure. We've ground the various forms of ochre into a powder to decorate our body and colour our hair, and made it into a paint for rock and cave art from at least 30,000 years ago. And it appears that our human species wasn't the first to have used these natural colours: ochre has also been found alongside flint artefacts in Neanderthal sites dating back over 200,000 years.[30]

What was truly transformative in the history of civilisation, however, was when we learned to extract pure metallic iron from these rust-coloured oxide ores. As we have seen, although there were a number of copper sources, tin was very scarce throughout the Bronze Age. Iron, on the other hand, is available in huge deposits and widely distributed around the world. But the reason that iron came to be exploited later than copper and bronze is that it is so much harder to tease this metal from its rocky ore.

The first furnace developed to smelt iron was the bloomer, where iron ore and charcoal were fired together but not at temperatures high enough to cause the iron to melt and flow away from the slag. Instead, the hot but still solid, spongy lumps – or 'blooms' – of iron mixed with slag were removed from the furnace and then hammered to separate the metal as pure wrought iron. 'Wrought', the archaic past particle of the verb 'to work', is an appropriate term: it takes an enormous amount of back-breaking labour with hammer and anvil to refine a bloom into pure iron. Iron smelting and smithing in this way was established in Anatolia by around 1300 BC.

A later development was to build much taller furnaces and pump a stream of air up from the bottom with bellows, to attain higher temperatures for melting the iron. This is the blast furnace. Adding limestone as a 'flux' helps the slag to flow, improving the separation of the iron and removing impurities. The molten metal can then be drained out of the base of the furnace as pig iron, or cast iron. Cast iron is high in carbon (around 3 per cent), which makes it strong but brittle. The first blast furnaces were operated as early as the fifth century BC by the Chinese, who in the first century AD were also the first to drive the bellows with waterwheels.[31] Blast furnaces and cast iron were adopted by the Arabs in the eleventh century, but didn't arrive in Europe until the late 1300s.[32]

Beginning at different times in regions around the world, the Iron Age transformed society. Bronze had remained relatively expensive, and so to a large extent it was the preserve of the ruling elites, or was used to equip the armies they pitched against each other. Iron ores, on the other hand, are plentiful and offer a general purpose metal for a whole range of practical artefacts. Iron implements are also much more durable and better at holding a sharp edge than those made of bronze. This was important not just for weapons and armour, but also for everyday tools. Iron axes made a huge difference for clearing forests to open up new areas of farmland. And iron-tipped ploughs not only increased the productivity of existing agriculture but enabled humanity to transform into fields land that was previously uncultivable. Both these tools opened up whole new regions for settlement.

In particular, the development from the late third century AD of the heavy mouldboard plough, with an iron cutting blade at the front of the ploughshare, made productive agriculture possible in the dense soils of the European landscape north of the Alps. Rather than just scratching a groove into the soil, the heavy plough slices deep into the sod and then flips it over round the curved mouldboard. The effect is to essentially turn upside down the entire topsoil, helping with weed control and mixing in fertiliser, and the furrows also greatly improve drainage of clay soils prone to waterlogging.[33] With this iron-made innovation the dense clay soils of Northern Europe became far more productive than the sandy soils around the Mediterranean. So with the help of iron axe and plough, the rolling North European plains were gradually transformed from post-Ice Age forests and waterlogged meadows into a great swathe of grain fields.[34] This in turn drove a fundamental shift in the population distribution and urbanisation of Europe over the subsequent centuries.[35]

If the material properties of copper are improved by mixing it into an alloy, the same is true of iron. Steel is an alloy of iron with a small amount of carbon, typically 1 per cent or less, and therefore sits midway in carbon content between pure wrought iron and cast iron. And like bronze, steel alloy is far harder than the pure metal. The exact properties of the steel can be tuned by varying the content of carbon: from soft but tough low-carbon steel to hard but brittle high-carbon steel. Over the centuries, metalworkers have developed a variety of techniques to achieve the desired amount of carbon: cooking wrought iron with charcoal so that it absorbs a little more carbon, or mixing proportions of wrought and cast iron. But high-quality steel remained laborious to create and so was reserved for critical applications like the cutting edge of knives and swords, or where its flexibility was required in small components, such as the springs in clocks.

Our current age of cheap, mass-produced steel began in the 1850s with the development of a simple way to remove carbon from pig iron. The Bessemer Process involves holding the molten pig iron in a tall cauldron and then blowing air up through the liquid metal. This burns off the carbon and removes other impurities, essentially creating a blank slate of pure iron, so that measured amounts of carbon can be mixed back in to produce whatever grade of steel you require. This innovation slashed the time to process 5 tonnes of steel from one day to about quarter of an hour,[36] prompting an explosion in steel output and dramatically dropping its cost. Thus the late Industrial Revolution transformed society into a far more metallic world. Today, steel is ubiquitous in household utensils and appliances, tools, machinery, railway tracks, ships and cars. We've also come to use it as the structural skeleton of our buildings, to provide embedded rebars to reinforce concrete and the framework of skyscrapers.

So if the Iron Age revolutionised human settlements, agriculture and warfare, our modern world is built with its alloy, steel. But where did this iron come from?

THE IRON HEART OF STARS

Ultimately, all the iron on Earth – from that in crustal rocks to the red-coloured haemoglobin carrying oxygen around in your veins – comes from the nuclear fusion reactions in the cores of stars. The universe created by the Big Bang contained mainly the simplest element, hydrogen, with some helium and a tiny amount of lithium thrown in. All the other elements in our periodic table were made by nuclear fusion in stars – cooked within their cores as they burned, or created when massive stars exploded at the end of their lifetime.

Iron is the star-killer element. When enough helium 'ash' produced by hydrogen fusion has built up in the core of massive stars, this then reacts to create heavier elements like carbon, oxygen, sulphur, silicon, and finally nickel and iron. Iron is the stablest element, and no more energy can be released by fusing it. As the giant star can no longer produce enough energy to hold up its outer layers, it collapses in on its own core, before exploding in an extremely powerful event known as a supernova. This final burst of fusion creates many of the heavier elements of the periodic table, and scatters all these atoms out into the cosmos. Several other key elements are produced by the violent collision of neutron stars, such as gold in a wedding ring, rare earth metals in a smartphone, lead on a church roof and uranium in a nuclear power station.[37] In this way, not only our planet but also the molecules of our bodies are made of stardust.[38]

The Earth formed out of a disc of dust and gas swirling around the proto-Sun around 4.5 billion years ago. Dust motes stuck together to build grains, which coalesced into larger and larger lumps of rock, and these accreted with gravity to form our planet. The heat of all these impacts melted the primordial Earth, and most of the dense iron sank to the very core, leaving a thick layer of silicate-rich mantle, which slowly cooled and solidified on top to form a thin crust. Many other metals dissolve readily in iron – they are known as siderophile ('iron-loving') – and so were also scrubbed out of the Earth's mantle and dragged down to the core as the iron sank. Consequently, siderophile elements like gold, silver, nickel and tungsten, as well as the platinum group metals we'll come to shortly, are depleted in the rocks of the Earth's crust. The precious gold that we've coveted through history was delivered to the Earth's surface by asteroid impacts after the planet had differentiated into its iron core and silicate mantle.[39]*

---

* The value that humanity has ascribed to gold through the ages is not just due to its rarity in the crust. Gold is unreactive, and so occurs as native metal – it is not bonded with other atoms in an ore – and its seams can be seen glinting out of a rock face or as flecks eroded out and redeposited in a river bed. This also means that it doesn't tarnish – its lustrous gleam doesn't dull; gold jewellery doesn't react with the moisture of your skin, nor do gold coins corrode away: they are stable stores of wealth. While other metals have a plain, hueless, silvery shine, gold is also special for its distinctive colour. Both its noble unreactivity and colour are in fact effects of Einstein's relativity. The outermost electron of the gold atom is moving at a fair fraction of the speed of light, and so owing to relativity becomes more massive and is pulled closer in to the nucleus. This both reduces its availability for chemical reactions and causes it to absorb blue light, so reflecting red and green to give a warm, golden colour.[40]

The iron heart of our world also serves to generate the Earth's magnetic field. Churning currents of molten iron in the outer core of our planet generate this field just like a dynamo. This has been hugely important since the eleventh century for the navigation compass used first by Chinese and then Islamic and European sailors (and for the migrating animals that were able to sense Earth's magnetic field long before us). But even more fundamentally, this magnetic cocoon has acted like a deflector shield to fend off the stream of particles gusting from the sun – called the solar wind – and so protect the Earth's atmosphere from being blown away into space. Thus the existence of complex life on Earth is itself dependent on this core of hot iron: the iron in your blood not only links you to the ancient stars that created it in their nuclear forge but also to the magnetic shield around our world that protects life on Earth.

Not all the Earth's iron sank to the core, though: it is still the fourth most abundant element in the crust, making up 5 per cent of the weight of all rocks on average. But to be useful for humanity, iron must have become concentrated into rich ores that can be mined and smelted. And this takes us to a very particular moment in our planet's history.

WHEN THE WORLD RUSTED

Virtually all the iron ore mined around the world throughout history comes from a kind of rock that formed during a singular period in the Earth's development.

Banded Iron Formations or BIFs (and the deposits eroded from them) make up by far the largest share of the iron ore we use. Each formation can be hundreds of kilometres long and several hundred metres thick, and the best ores contain more than 65 per cent iron.[41] As their name implies, they have a distinctive stripy appearance, each band being between a millimetre and several centimetres thick. The layers are made up of iron-oxide ores – haematite and magnetite – alternating with chert or shale rock.

And they are almost unimaginably old. The great majority of Banded Iron Formations were laid down in a relatively brief period of worldwide deposition 2.2–2.6 billion years ago,[42] around the time the first continents were forming on our planet.* The fact that iron ore all over the world dates to pretty much the same moment in the Earth's history indicates that something truly profound was happening to the planet at that point. The BIFs were laid down on the bed of the ancient oceans, and their stripes reveal fluctuating conditions in the primordial waters: the ore sedimented as a gentle drizzle of grains of iron minerals settled out of the water onto the sea floor, alternating with periods of deposition of normal marine mud. But the curious thing is that today iron only dissolves in seawater in vanishingly tiny concentrations. So how was all this iron deposited from the seas in a prolific spell around 2.4 billion years ago? What was different back then?

If you were to travel back to this time of BIF, you'd encounter a truly alien world. The young Earth was still much hotter inside than today, and this would have driven rampant volcanism. The planet-spanning ocean was broken only by arcs of volcanic islands and tiny continents that had begun to emerge. Ultraviolet radiation from the sun blazed down onto the barren surface. The skies were probably permanently shrouded in sickly-yellow, hazy clouds and the air was full of nitrogen and carbon dioxide. And, crucially, there was no oxygen – you would have needed a spacesuit to walk around on your own homeworld.

Today, oxygen makes up a full fifth of every breath you take. But for the first half of the Earth's lifetime, the world had essentially no oxygen gas in its atmosphere and oceans. The

---

* A second, minor spurt of BIF occurred around 1.8 billion years ago, producing the Gunflint and Rove Formations stretching between Minnesota and Ontario alongside Lake Superior.[43]

oxygen in our air, and that dissolved in the seawater, was put there by life. Some organisms are able to harvest the energy in sunlight to transform carbon dioxide into the organic molecules that make up cells, and in the process they split water, H_2O, to release the oxygen as a waste gas. This biological alchemy is known as photosynthesis, and it empowers the cell to be incredibly self-sufficient and manufacture all it needs from only light, carbon dioxide and a few other dissolved nutrients.

The kind of cells that developed this ability to photosynthesise and release oxygen are known as cyanobacteria.[44] All the more complex sunbathing life forms – diatoms, algae, seaweed as well as all plants and trees on land – inherited this capability from a crucial evolutionary event about a billion years ago when their single-celled ancestor took cyanobacteria inside itself. And it was these minuscule early cyanobacteria, swarming in the primordial seas and giving off oxygen exhaust fumes from their photosynthetic machinery, that eventually oxygenated the entire planet. Geologists studying the change in ancient rocks can see a sharp indicator of the first rise in oxygen levels 2.42 billion years ago, known as the Great Oxidation Event (GOE). Although this only saw oxygen levels rise to perhaps a few per cent of today's,[45] still far too low for a breathing human, it had profound implications for the chemistry of the Earth and the development of life. In fact, the GOE is the most significant revolution in the history of the planet.[46]

Shortly after the Great Oxidation, around 2.2–2.3 billion years ago, the Earth appears to have descended into the longest, and probably the most severe, instance of glaciation in our planet's history. Back then, the sun was about 25 per cent dimmer than it is today, and to remain warm enough for water on its surface to remain liquid the Earth would have needed a substantial greenhouse effect to insulate the world. The ancient atmosphere contained significant amounts of methane, which is a powerful greenhouse gas, but the increased oxygen would have reacted with the methane and removed it, effectively stripping the planet of its warming blanket. Temperatures plummeted and caused a global glaciation, dubbed Snowball Earth, with thick ice smothering almost the entire surface of our planet.[47] It remained locked in this whitened state for 10 million years[48] until the ongoing volcanic activity had built up enough carbon dioxide in the atmosphere for the big thaw to begin. Rescuing the planet from such deep glaciations is one of the major benefits of volcanism for life on Earth.[*]

Many microorganisms that were around at the time of the Great Oxidation Event could not cope with reactive oxygen gas and were wiped out by this toxic pollution – effectively an oxygen holocaust. In order to survive in the new world order, organisms had either to evolve to survive the presence of this toxic gas – by developing ways to exploit its reactivity to unleash greater amounts of energy from their metabolism, as our cellular ancestors did – or else become restricted to secluded habitats where oxygen doesn't penetrate, like sea-floor mud or deep underground.[†]

But more complex multicellular life like animals and plants depends on oxygen to survive, as well as an ozone layer to shield the planet's surface from destructive UV rays. And so although there were vast numbers of organisms poisoned by the reactive oxygen gas or banished to anoxic refuges, the

---

[*] Before oxygen started to build up in the air, the atmosphere also contained no ozone layer which is itself formed from oxygen high in the atmosphere, and so harmful ultraviolet radiation from the sun would have streamed down to the surface of our planet. This high-energy light would also have driven chemical reactions within the atmosphere to create tiny droplets of hydrocarbons, enshrouding the early Earth in a smoggy photochemical haze. But the accumulation of oxygen in the air reacted with this yellowy haze to scrub it clean – and the skies turned blue.

[†] When animals evolved much later, they provided new anoxic refuges within their own bodies. The oxygen-free gut of a ruminant like the cow recreates a small pocket of the primordial Earth, allowing anaerobic microbes to flourish and run their ancient metabolism that produces methane, which the cow then releases from both ends.

Great Oxidation Event paved the way for all complex life on the planet. Atmospheric levels finally approached those of today, sufficient for the emergence of animal life, around 600 million years ago.

This brings us back to the creation of the Banded Iron Formations that we mine around the world. Oxidised iron is barely soluble in water – and this explains why in today's well-oxygenated oceans iron is so scarce. But the reduced form of iron dissolves very well, and so on the primordial Earth before the Great Oxidation Event levels of this reduced, soluble form of iron were very high in the oceans, released from submarine volcanoes or washed in by rivers from the eroding landmasses. During the Great Oxidation, cyanobacteria proliferating in the oceans slowly but surely oxygenated the surface waters. The ocean depths, however, remained anoxic and so were rich in dissolved iron – around 2,000 times more than what we find in the seas today. But every time deep water was pushed up onto shallow marine shelves it mixed with oxygen, the iron was oxidised so that it could no longer remain dissolved, and it settled onto the sea floor, creating the Banded Iron Formations. And so the planet rusted.

Virtually all the iron ore mined today and throughout history was created within 200 million years of the Great Oxidation Event 2.42 billion years ago as BIFs. In this way, today's blue skies, the lungfuls of life-giving air we inhale, and the iron that has provided the tools of our civilisations for millennia are all deeply linked. And oxygen has another benefit: it enables us to make use of fire.

For 90 per cent of the planet's history there has been no fire on Earth. While there were volcanic eruptions, there was not enough oxygen in the atmosphere to sustain combustion.* Thus the rise in oxygen not only allowed more complex life to evolve

---

* The first signs of charcoal in the fossil record, an indicator for wildfires, don't appear until around 420 million years ago, when atmospheric oxygen levels first rose above 13 per cent.[49]

on Earth, but it gave humanity fire as a tool. We first used it for keeping the night's cold and predators at bay, cooking food, and clearing land. Humanity then learned how to exploit the transformative heat of fire: to bake clay into hard ceramic pottery or building bricks, to make glass, or to smelt metals for tools. Today, we use fire for generating electricity and driving a huge range of industrial processes; and we harness tiny bursts of flame in the engine cylinders of our cars. We are as utterly reliant on fire today as were our Palaeolithic ancestors who huddled around a campfire; we've just hidden it behind the scenes of the modern world.

THE PERIODIC TABLE IN YOUR POCKET

In the ancient world, only a handful of different metals were used across society, including copper and zinc in bronze utensils, iron in steel tools and weapons, lead in plumbing, and precious metals like gold and silver in decoration, jewellery and currency. These metals remain important in the modern world and, indeed, we still very much live in the Iron Age. Iron, and especially that mixed into the alloy steel, accounts for around 95 per cent of all metal used by today's industrialised civilisation. Other metals are still crucial, but the applications we put them to have shifted substantially. Copper, for example, was used first as a major alloy component for the tools and weapons of the Bronze Age, but it declined in significance and trading value with the development of iron smelting and the availability of this superior metal. But in the last two centuries copper has resurged in importance as a relatively abundant metal that conducts electric current well, providing the wiring of our modern electrified world. We are using the same Bronze Age metal, but reflecting technological change through history, we now exploit different properties.

We've also discovered and learned how to make use of new metals. One of the most prominent is aluminium. This is in fact the most abundant metal in the Earth's crust (about 8 percent overall), but it is devilishly difficult to separate from its rocky ores. It wasn't until the end of the nineteenth century that we learned how to mass-produce it cheaply, by passing electricity through its molten ore. It then became widely used as a building material and for food packaging. In particular, aluminium is very lightweight, and so it came into its own with the expansion of aviation from the First World War. But it is in recent decades that the number of metals we use in our technological society has really exploded.

How many different kinds of metal do you think you have on your person right now? A handful? A dozen? You may be astonished to hear that today over 60 different metals are employed in a single hand-held electronic device alone. These include base metals such as copper, nickel and tin; special-purpose metals like cobalt, indium and antimony; and the precious metals gold, silver and palladium.[50] Each one is exploited for its particular electronic properties, or for the tiny, powerful magnets used in the speaker and vibration motor. A whole range of non-metal elements are included in your smartphone too, such as carbon, hydrogen and oxygen in the plastics, bromine as a flame retardant, and silicon for the microchip wafers. Of the 83 stable (non-radioactive) elements in existence, around 70 are used in making an everyday consumer device like a smartphone[51] – which means you carry about 85 per cent of the entire available terrain of the periodic table in your pocket.

It's not just electronics that employ such a multitude of metals. The high-performance alloys used in the turbines of a power station or an aircraft jet engine mix more than a dozen, and the reaction-accelerating catalysts in the chemical industry – including those making modern medicinal drugs – employ more than 70 different metals. Yet most of us have never even heard of many of these critical metals – elements with exotic names like tantalum, yttrium or dysprosium.

This expansion in the diversity of metals we've come to exploit has been staggering. While microchips today contain around 60 different metals, as recently as the 1990s it was only about 20.[52] Take indium, for example. This metal was discovered in 1863, and in the Second World War was used to coat bearings in aircraft engines to protect them against corrosion. But it wasn't until the 1990s that indium came into widespread use when a thin film of indium-tin oxide was added to our screens, exploiting a rare combination of properties – the metal oxide is both transparent and electrically conductive. Today indium is used in everything from flat-screen TVs to laptops, and in particular the touch-sensitive screens of modern smartphones and tablets.[53] Similarly, gallium was discovered within a few years of indium, but again didn't find any widespread application until the electronic age: today it is used in integrated circuits, solar panels, blue LEDs and laser diodes for Blu-ray discs.

Most of these exotic-sounding metals belong to one of two groups: the rare earth metals (REMs) and the platinum group metals (PGMs). The metals in each of these two sets are chemically very similar, which means that they have become concentrated in the same minerals and are extracted at the same time by our separation processes. These roughly two dozen metals really do define our present technological age – over 80 per cent of their exploitation has happened just since 1980.[54] And if they are the key ingredients of our current technological age, they will be even more crucial in the future as we transition away from the current carbon economy. They will give us the compact but powerful magnets needed in the generators of wind turbines and the motors of electric vehicles, as well as high-capacity rechargeable batteries.

The seventeen rare earth metals are made up of the 'lanthanide' series of elements in the sixth row of the periodic table, as well as the chemically similar elements scandium and yttrium. Their title is something of a misnomer, though, as they're not actually all that rare in the planet's rocks – apart from the radioactive promethium, of which there is no more than about half a kilogramme in the entire Earth's crust.[55] Lanthanum, for example, is almost as abundant as copper and nickel, and in fact three times more so than lead. And all the REMs are at least 200 times more common than gold.

So the problem isn't so much their overall abundance in the crust, but the difficulty in extracting them. The fact that the rare earth metals are chemically similar and so occur in the same kinds of minerals means that they are also difficult to isolate from each other as pure metals. Even more troublesome is the maximum concentrations at which they occur within rocks. Many other metals become concentrated by particular geological processes into rich ores, such as the Banded Iron Formations or the thick seams of silver running through Cerro Rico that we'll come to in Chapter 8. But the chemistry of the REMs means that they tend not to become enriched into high-grade ores, but instead are mostly thinly dispersed in low concentrations through rocks. On the whole, therefore, mining them specifically isn't economically feasible – it costs more to extract them than they are worth. Thus the geographic avail-ability of rare earth metals that can be mined profitably is limited around the world. They are extracted today in small amounts in India and South Africa, but since the 1990s the vast majority of global production has been in China.

The six platinum group metals – rhodium, ruthenium, palla-dium, osmium, iridium and platinum – are clustered in the middle of the periodic table, and like the REMs they are chem-ically similar, which again means that they tend to occur together in the same mineral deposits. But unlike their rare earth cousins, the platinum group genuinely *are* precious metals. They are among the rarest stable elements in the Earth's crust – some are millions of times scarcer than copper. Platinum itself is one of the more common metals within this group, yet worldwide production is only a few hundred tonnes a year, compared to 58 million tonnes of aluminium or over 1 billion tonnes of pig iron. Iridium is particularly rare and present in the Earth's crust at only about one part per billion: on average, 1,000 tonnes of crustal rock contain no more than 1 gramme of iridium. Like the other platinum group metals (and gold), iridium is a sidero-philic element and consequently virtually all that was present on the primordial Earth was dragged deep into the interior as iron sank inwards to form our planet's core.*

The PGMs are also known as noble metals, as they are resistant to chemical attack and corrosion, even at high tempera-tures. Being both rare and unreactive, platinum is an attractive material for jewellery, and about a third of the annual produc-tion of this precious metal goes to adorning our bodies.† But unlike other precious metals like gold – which is primarily used today in jewellery or reserves of wealth, with only about 10 per cent going to industry, mainly as electrical contacts – the platinum group metals are employed in a huge range of prac-tical applications: they are used in everything from turbine engines to spark plugs, from computer circuits and hard drives to contacts in heart pacemakers.[57]

The majority of platinum itself is used for catalytic converters on vehicle exhausts to reduce harmful emissions, and for catalysts for the chemical industry. These are employed for refining petrol-eum and creating pharmaceuticals, antibiotics and vitamins, as well as in the production of plastics and synthetic rubber. Perhaps the most significant use, however, is for agriculture. Here it serves as a catalyst in the chemical process that produces artificial fertiliser – an activity that effectively mines the atmosphere for

* However, iridium is over a thousand times more common in asteroids, which were too small to undergo this differentiation process into iron core and silicon-rich mantle and crust. High concentrations of iridium in the thin clay layer around the world that marks the boundary between the Cretaceous and Palaeogene geological periods is therefore one of the strongest pieces of evidence that an asteroid or comet impacted the Earth 66 million years ago, at the time of the 'dinosaur-killer' mass extinction.

† The name platinum comes from the Spanish for 'little silver'. Platinum has a long history of being formed into ornaments by pre-Columbian South American natives – the metal can be found among river-bed sand in Ecuador and Colombia – before it was first brought back to Europe by a Spanish military commander.[56]

nitrogen.[58] It is estimated that today around half of the human population are fed with the help of this metal.[59]

The extreme rarity of platinum group metals means that they can only be mined from rocks where these elements appear in substantially higher concentration than their average in the Earth's crust. They are therefore limited to sites that have undergone somewhat quirky geological processes. The platinum group can become enriched within certain ores of copper and nickel, and so some PGM extraction is achieved as a by-product of mining these industrially significant metals. Sources include mines near Norilsk in Russia, where deposits formed by the eruption of the Siberian Traps at the end of the Permian Period 250 million years ago (see pages 141–2) are being dug,[60] and the Sudbury Basin in Canada. The Sudbury Basin is one of the largest, and oldest, impact craters known on Earth. This crater was originally around 250 kilometres across, and formed 1.85 billion years ago when an asteroid over 10 kilometres in diameter slammed into the planet. This colossal hole in the ground filled with magma containing copper, nickel, gold and platinum group metals, which then crystallised into rich ores.[61] But by far the greatest source of platinum group metals in the world is a single region of South Africa.[62] Around 95 per cent of the global reserves of PGMs occur in what is known as the Bushveld Complex.[63]

The Bushveld Complex is one of the most metal-rich spots in the world. This is a vast saucer-shaped lump of igneous rock, about 450 by 350 kilometres in size and up to 9 kilometres thick in places. It formed about 2 billion years ago – not too long after the Banded Iron Formations were deposited in the oceans around the world – when a huge mass of magma intruded to within a few kilometres of the surface, and then cooled slowly underground. As the magma cooled, different minerals separated out and solidified, like a huge layer cake. One of these layers was enriched in platinum group metals to a level of about ten parts per million, substantially higher than most other rocks, but still offering only about 5 grammes of platinum and palladium for every tonne mined.[64] It's still not entirely clear what unusual geological conditions acted to concentrate such rare PGMs about a thousandfold, but 2 billion years later it is this thin layer that we now mine for the vast majority of the platinum group we use.[65]

Historically, metals have been used for their mechanical strength for tools and weapons. Today we still employ a wide range of them in construction, and high-performance alloys serve in power generation, transport and industry. But we've also come to use a staggering diversity of metals for their catalytic properties in accelerating chemical reactions – which, as we saw, includes helping feed the global population – or their electronic characteristics for modern devices. Compared to the metals of antiquity like copper or iron, many of these elements that run the modern world are very hard to find in appreciable ores around the world, and the Earth has only provided us with them in rare spots with unusual geological conditions. Indeed, several of the metals we've looked at in this section are now considered as 'endangered elements' of the periodic table.

ENDANGERED ELEMENTS

One of the biggest concerns for continuing to meet our industrialised world's appetite for resources is the future availability of several of the most important technological metals. Endangered elements include some of the PGMs, several REMs, and lithium, the lightest metal, used in rechargeable batteries. Indium and gallium too are among those picked out as being under serious threat in the coming years.[66]*

---

* Although not a metal, helium is also highlighted as being critically endangered. Helium is used not only for filling party balloons, but in ultra-cold liquid form serves to chill the superconducting magnets used in MRI scanners in hospitals, or in science labs around the world. Helium is actually the second most abundant element in the universe, but because it is such a light gas its atoms readily escape from the Earth's atmosphere and into space (whereas the powerful gravity of the gas giant planets Jupiter and

The problem isn't that these elements will disappear altogether, but that the rising demand for technological applications could greatly outstrip their limited supply. Take the rare earth metals, for example. That the world has become so reliant on Chinese production of the REMs – currently around 95 per cent of the global total – causes a great deal of concern over ensuring that their supply continues to meet the growing demand. This is only heightened by the fact that in many cases there is no known alternative metal that performs just the right function. REM prices spiked in 2010 after China announced a 40 per cent cut in its export quota, citing its own domestic demand and environmental considerations. Although this has been relaxed again there is still great concern over the continued supply of these elements so critical to our technologies.[67]

As is normal when supply restrictions cause price rises, this created the economic incentive for other sources to be exploited, and new mines and refining facilities are opening up in Australia, Brazil and the United States. But even when these become fully operational China will still dominate the production of the heavy REMs, which are the scarcest and most valuable of the rare earths.[68]

But another, far more surprising solution is being considered. Some of the scarce metals used in modern electronics, such as the indium of your smartphone's touchscreen, are used in vanishingly thin films or mixed in tiny quantities with other metals, which makes them hard to recycle at the end of the

Saturn holds on to a substantial fraction of helium in their atmospheres). Helium on Earth is produced deep underground. When radioactive elements like uranium decay they release a form of radiation called alpha particles, which are just the nuclei of helium atoms. This helium becomes trapped underground by the same geological conditions as natural gas (itself formed in the same process as oil, as we'll see in Chapter 9), and so most helium is commercially extracted from natural gas production. In this way, not only is helium gas mined from deep underground, but the floating balloons at a children's birthday party are filled with atoms that were once fast-moving radiation particles.

device's lifetime. Many others, however, can be recovered with a little effort. After decades of our simply discarding obsolescent gadgets, many landfills may now hold veritable mother lodes of these valuable metals. And this raises an intriguing possibility: landfill mining – picking back over our rubbish for the buried treasure it contains. A test site at a landfill 60 miles east of Brussels, for example, aims to recover building materials and convert waste into fuel, but also seeks to sort and recover valuable metals. And landfill mining could begin soon in Britain too: four sites that have been tested were found to hold significant amounts of aluminium, copper and lithium.[69] The opportunities for prospectors are particularly good in the high-tech dumps of Japan, however. It has been calculated that its buried waste contains three times the global annual consumption of gold, silver and indium, and perhaps as much as six times that of platinum. In fact such artificial ores made up of reduced mobile phones can contain thirty times the concentration of gold as an actual goldmine.[70]*

This chapter has brought us from the Bronze Age to the modern world of high-tech metals, and explored how particular geological conditions on our dynamic Earth provided us with raw materials for the tools of civilisations. But precious metals like gold and silver have also served through history as a medium of exchange – they were minted into coins to facilitate commerce and trade between disparate cultures. One of the earliest long-range overland trade networks stretched across Eurasia and connected China and the Mediterranean: the Silk Road.

---

* Another particularly intriguing suggestion has been put forward for supplementing the supply of platinum group metals from naturally forming rocks. The lighter PGM elements – ruthenium, rhodium and palladium – are created in significant amounts as by-products from the atom-splitting of uranium in nuclear reactors, and could be economically extracted from the spent fuel rods. This represents real-life alchemy – transmuting one element into another – although not through the discovery of a philosopher's stone but using means that would be beyond the comprehension of the alchemists of history: the atom-morphing reactions of nuclear fission.[71]

Chapter 7

Silk Roads and Steppe Peoples

The continent of Eurasia, stretching 12,000 kilometres from the Atlantic to the Pacific oceans, contains over a third of the total land surface area of our planet, and has hosted many of the most sophisticated civilisations in history. It is in Eurasia that different cultures developed wheeled transport, iron-smelting, transoceanic trade links and industrialisation. Two aspects have defined the course of history across this sprawling land mass: long-distance trading routes over the great breadth of the continent, and nomadic peoples repeatedly spilling out of the continental interior to challenge the civilisations growing around its margins. It is the fundamental planetary characteristics of climate bands, and the environments within them, that have created these themes.

THE HIGHWAY ACROSS

Long-distance overland trade across central Eurasia was well established by the first millennium BC to satisfy the Chinese demand for jade from Central Asia and the Mesopotamian desire for lapis lazuli from Afghanistan.[1] But this long-range commerce intensified dramatically from the first century AD. By that time two great powers had arisen on opposite sides of the wide Eurasian landmass: Han China to the east and the Roman Empire in the west.

In China, civilisation had begun along the riverbanks of the River Wei and lower Yellow River,[2] before spreading further south to the Yangtze. It is this plain between the mighty Yellow and Yangtze rivers that forms the heartland of China.[3] Wheat and millet were grown in the drier north, and rice in the wetter climatic zone of the south, where two crops could be harvested every year.[4] Whereas the fields of Egypt were rejuvenated annually by the flooding of the Nile, Chinese farmers had received their endowment of fertile soils as a lump-sum deposit. Blankets of loess soil were formed over the last 2.6 million years of the recurring ice ages by windblown dust from retreating glaciers and desert regions.[5] Accumulations of this fertile soil can be 100 metres thick in places, forming impressive plateaus, but it is also eroded and deposited by rivers in alluvial plains.[6] Loess soil is mineral-rich, porous and has a distinctive buff colour – indeed, the Yellow River is named after the loess sediments that it carries.*

This agricultural core of modern China was unified in 221 BC, after 250 years of warfare, by the victorious Qin dynasty (which gave us the name China). Like Egypt, China was able to achieve such early and long-lasting political unification, and protection from external threats, because of its natural frontiers:[8] the Pacific coastline to the east, the inhospitable highlands of the Tibetan Plateau and Himalayas to the west, and dense jungle to the south. The main weakness was the northern boundary, marked not by a distinct topographic feature like a mountain range but by a smooth ecological gradation from the fertile agricultural plains into the Gobi Desert and then the arid grasslands of Central Asia. By around AD 100, under the Han dynasty, the Chinese empire had expanded north to the Gobi Desert and Korean Peninsula. It also reached west in a long arm following the contours of the landscape through the Gansu Corridor, marked

---

* Loess soils cover no more than 10 per cent of the Earth's surface, but offer some of the most agriculturally productive land in the world. Alongside the thick loess plateau in China, a wide band of loess runs through the steppe region of Central Asia, and there are also patches of these fertile soils across Northern Europe.[7]

by a string of oases between the towering Tibetan Plateau and the Gobi Desert, and into the Tarim basin holding the Taklamakan desert to protect its trade routes across Central Asia.

The expanse of the Roman Empire too was defined by natural boundaries. By AD 117, the time of its greatest extent, Rome had expanded from a small town halfway up the Italian Peninsula to a vast empire encompassing around a fifth of the global population at the time. At this high-tide mark, the Roman Empire completely encircled the Mediterranean – or *mare nostrum*, 'our sea', as it was called – its frontiers following the features of the landscape. In the west, the empire stretched to the Atlantic coastline of the Iberian Peninsula and Gaul (France), and up through drizzle-swept Britain. Its northern limits rested along the banks of the Rhine and Danube rivers that snake through the European plains. The frontier followed the Carpathian Mountains to the shores of the Black Sea, and then along the line of the Caucasus. Reaching down through Mesopotamia and round the coastline of Palestine, the empire then extended along the Nile, and finally followed the North African coast until the land gave way to the inhospitable dust of the desert.[*]

---

[*] This extent of the Roman Empire has had a long-lingering influence through history, and its imprint is still evident in the geographical spread of the three forms of the Christian faith in modern Europe – Catholicism, Protestantism and the Eastern Orthodox Church. The East–West Schism of 1054 saw Christianity divide into two main branches: Roman Catholics led by the Pope and Eastern Orthodox Christians headed by the Patriarch in Constantinople. The second major schism was the split of Protestantism from Catholicism in the sixteenth century, the result of the Reformation that began in Germany (territory that had remained outside the Roman Empire). This tripartite division of Europe largely separated along two main fault lines. The first, between Catholicism and Eastern Orthodoxy, tracked the Danube river as it ran south through the plains of Hungary: the old demarcation of the zones of influence of the East and West Roman Empires, roughly equidistant between their capitals in Rome and Constantinople. The second line follows the long-standing frontier of the Roman Empire along the Rhine, and the boundary between the Latin civilisation and the Germanic tribes, with the territories that embraced Protestantism lying beyond the old Roman border. In broad brushstrokes, the three Christianities follow the frontiers of the old empire, themselves defined by natural boundaries of the underlying landscape.[9]

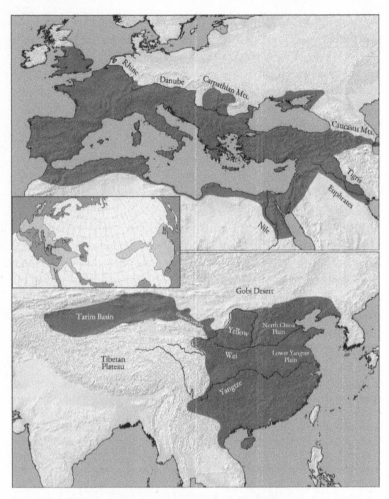

The limits of the Roman (*top*) and Han Chinese (*bottom*) empires in the second century AD were defined by natural features.

At the beginning of the second century AD, the Roman and Han Chinese empires shared many features. Both had roughly the same population of about 50 million, and covered approximately the same territorial area – around 4–5 million square kilometres. The Roman Empire was based around the rim of its internal sea, the Mediterranean, which made for easy internal transport and trade, whereas the core of China spread across plains watered by the mighty Yellow and Yangtze rivers. Rome built roads for overland transport, China more canals, and both civilisations constructed fortified walls to keep the barbarians at bay.[10]

At this point of their greatest extent, the Roman and Han territories combined stretched to encompass a full three-quarters of the complete breadth of the Eurasian continent between the Atlantic and East China Sea. And they were brought together by the trade of one precious commodity – silk.

China had been using silk to pay off the aggressive Xiongnu tribe beyond its northern borders, or to buy their horses,[11] and it already traded silk to Persia. But now it found an eager new market even further afield with Rome, where the elites valued this beautiful fabric from the East.[12] Chinese silk first reached the eastern Mediterranean by overland caravans,[13] but was also traded along the sea passage we explored in Chapter 4: by ship across the Indian Ocean, up the Red Sea, through the desert by camel to the Nile and then by boat to Alexandria.[14]*

Trade along the Roman–Han axis reached its peak in the early second century AD, before the collapse of the Han dynasty in AD 220 and the slow decline of the Roman Empire. But commerce between East and West continued through the centuries. Today we know this long-range trade between the

---

* The fact that silk arrived by these contrasting routes caused the Romans to believe that it came from two different places: transported by land from the country of the Seres, and reaching the west across the water from that of the Sinae.[15] The Romans were also unclear about how silk thread was produced, believing it was combed from leaves in the forest – a misunderstanding perhaps deriving from the fact that the caterpillars of the silk moth are fed on mulberry leaves.[16] The Han Chinese had a similar misapprehension about the natural source of the cotton they received from India, believing it to be the 'hair combed from certain water sheep'[17] and not, in fact, the fluffy fibres encasing the seeds of a plant that is related to okra and cocoa.

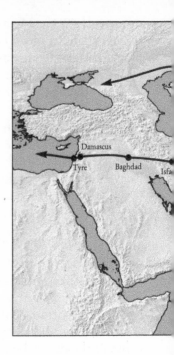

The main overland routes and entrêpots
of the Silk Road across Eurasia.

extremities of Eurasia as the Silk Road. But the term is a
misnomer. There was never just one road, but rather an
extensive network of routes linking cities, oasis towns and
trading entrepôts – an entire web of transport and commerce
draped across Central Asia. And although we usually imagine
the Silk Road as a transcontinental link between its remote
termini in China and the Mediterranean, trade in between
these waypoints was just as crucial, with the routes extending
into northern India and Arabia.

The history of the Silk Road illustrates the extraordinary
extent to which the terrain of our world has ordered and
directed our movements, lifestyles and trade. From the northern
plains of China the Silk Road passed along the Gansu Corridor,

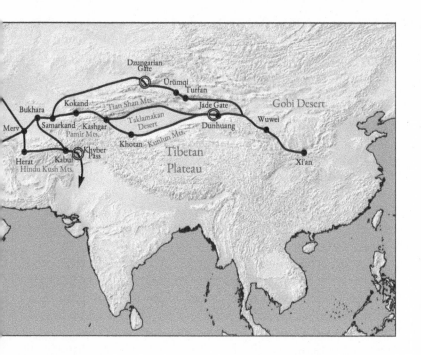

a 1,000-kilometre-long passage running between the towering Tibetan Plateau and the Gobi Desert. After passing the oasis city of Dunhuang and the Jade Gate of the Great Wall, the route reaches the lip of the Tarim Basin and the punishing Taklamakan desert lying in this depression. One fork of the Silk Road now headed north along the feet of the Tian Shan mountain range, while the other passage hugged the southern edge of the desert that meets the Tibetan Plateau. Both routes converged again at Kashgar and the road then threaded through mountain passes across either the Tian Shan to the west or the Pamir mountains to the south. A further route crossed through Ürümqi and the northern Tian Shan, making use of the Dzungarian Gate valley to pass through the mountains.

After negotiating the Taklamakan desert and Tian Shan mountains, the Silk Road passed along valleys and then wove across the deserts of Central Asia – through modern-day Uzbekistan, Turkmenistan and Afghanistan – connecting oases and trading stops like Samarkand, Bukhara, Merv and Herat. A southerly branch of the caravan network bore south to Kabul, and from there threaded through the Khyber Pass over the Hindu Kush mountains of the Western Himalayas and down into the Indus valley.[18] Continuing west, the Silk Road passed south of the Caspian Sea through Persia, linking large entrepôts like Baghdad and Isfahan, and then carried on to Damascus and the ports of the eastern Mediterranean; or it turned north to the Black Sea, from where the goods were carried to Europe by ship.

The exact nodes in this trans-Asian network varied over history, with consecutive empires routing the trade through their preferred cities, but this broad outline gives us a fair idea of the enormous, sprawling web of what we have come to call the 'Silk Road'. And much of the great east–west communications network across the breadth of Asia passed through a particular climate band – desert.

The particular environment of the Silk Road was dictated by invisible movements of the atmosphere high above the heads of the travelling merchants. Around the equator, where sun-driven evaporation and rising air causes lots of rainfall, we find the dense tropical rainforests of our planet, stretching across the Amazon, the islands of the East Indies, and central and western Africa. (As we saw in Chapter 1, East Africa's original rainforest was replaced by dry savannah due to the tectonic uplift of the Great Rift System.) But by the time this air has rolled over through high altitude and sunk back to the surface again, at around 30° north and south of the equator, it has become very dry: and here it creates the most arid regions on the planet's surface. In the Southern Hemisphere, this dry band includes the Great Sandy Desert of Australia, the Kalahari in South Africa, and the Patagonian Desert in South America.

In the mirror-image band stretching across the Northern Hemisphere lie the Mojave and Sonoran deserts in America, the Sahara, the Arabian Peninsula and the Thar Desert in north-west India.

The pattern is slightly more complicated in South East Asia, however. Here the desert band is disrupted by the monsoon system and its seasonal high rainfall. We'll see in Chapter 8 how the Tibetan Plateau and Himalayas act to strengthen the monsoons over India, but these high mountains, and their offshoots like the Pamir, Kunlun and Tian Shan ranges, also act to block moisture-laden air from the Indian and Pacific Oceans from entering Central Asia. Many of the deserts, such as the Gobi and Taklamakan that the Silk Road had to negotiate, are created by this rain-shadow effect, and as a result the desert band in Asia extends much further from the equator than on other continents. Although some of these deserts are beset with shifting sand dunes – the Taklamakan is the second largest shifting-sand desert in the world after the Rub' al-Khali that covers most of the southern Arabian Peninsula – many have a hard, pebble-strewn surface that is readily passable if you carry enough water supplies.

So not only did the crumpling tectonic uplift over the past 40–50 million years build the wide arc of the Himalayas but it also formed the deserts behind them. And it is both these deserts and the mountain ranges that defined the landscape the Silk Roads threaded across. Here one animal was uniquely suited to moving through this arid climate band and facilitate trade between east and west: the camel.

As we saw in Chapter 3, the camel evolved in North America and migrated across the Bering land bridge during an ice age several million years ago. While it died out in its birthplace, two varieties developed in the Old World: the two-humped bactrian camel in Asia (domesticated about 3000 BC)[19] and the single-humped dromedary in the hotter deserts of Africa (domesticated around the second millennium BC).[20] The camel's ability to carry much greater weight for longer before needing rest, and its need for much less water, made it far superior to the horse or donkey for transport through these arid regions.

Contrary to popular belief, camels do not store water in their humps, which is in fact a store of body fat. Rather than distributing fat all over their bodies in an insulating layer, as many mammals do, camels use their humps as fat reservoirs, which provide energy while allowing the animal to remain cool. The camel is uniquely adapted to desert survival. After a week or so of trekking through an arid landscape, it can have lost almost a third of its body water with no ill effect[21] – the animal can cope with such extreme dehydration without its blood becoming dangerously thick. The camel's kidneys and intestines are able to produce highly concentrated urine and dung so dry it can be used to fuel a fire; it can also recapture moisture it would otherwise have breathed out,[22] the water recondensing in its nasal passage like the drips from an air-conditioner unit. And the padded feet of the animal allow it to traverse such diverse terrain as desert sands, swamps or rock-strewn landscapes.[23]

Camels were essential for the trade in incense which began about 4,000 years ago. Although Arabia lies within the Earth's desert band, in the south-west of the peninsula mountains capture enough rainfall from the summer monsoons to create a rare patch of vegetation. Here frankincense and myrrh can be extracted from small shrubby trees growing among these mountains. As they are best harvested in spring and autumn, their growing cycle is out of synch with the seasonal monsoon winds that facilitate sea transport up the Red Sea to Egypt or across to India, and so overland travel by camel was far more suitable. Caravans carrying incense skirted up the coast of the Red Sea through the Arabian Desert and then across the Sinai to Egypt and the Mediterranean, or they made their way east to Mesopotamia.[24]

In North Africa, camel caravans crossed the Sahara from about AD 300,[25] bringing Sudanese gold to the Mediterranean.

On their return the traders carried common salt mined from beneath the sands (which had been deposited by vanishing lakes as the Sahara desiccated) south to the trading town of Timbuktu, where it was loaded onto canoes and taken by river deeper into Africa. By the early thirteenth century the Empire of Mali had emerged, fed by the belt of fertile soil alongside the River Niger and its tributaries and by mining rich gold fields, and Timbuktu now became a royal city.[26] The salt-for-gold trade persisted for centuries, and finding the mysterious source of the precious metal was one of the major motivations for Portuguese sailors exploring the West African coast in the early 1400s (to which we'll return in Chapter 8).

Camels were also critical to the Silk Road crossing the arid latitude zone of Asia. Here too the animals were ideally suited for traversing diverse landscapes: sure-footed on their padded hooves over rock-strewn ground, and able to tolerate the climatic extremes between the deserts and high mountain passes.[27] With a single pack camel able to carry a load of over 200 kilogrammes, and caravans easily numbering several thousand animals, their total cargo could rival that of a large merchant sailing ship.[28]

Although the arduous overland journey meant that in general it was high-value commodities that were moved along the Eurasian trade network, silk was not the only merchandise.* Spices like pepper, cinnamon, ginger and nutmeg were carried west. India traded cotton and pearls, Persia exported carpets and leather, Europe sent silver and linen. Rome traded its high-quality glass, and topaz and coral from the Red Sea. Frankincense from the south Arabian Peninsula, as well as precious stones and dyes such as indigo, were also carried across Central Asia.[30]

---

* Silk became less important along these east–west trade routes after around AD 550, when eggs of the silk moth were smuggled to Constantinople, launching a new silk industry that undermined the previous Chinese monopoly.[29]

But the huge importance of the Silk Road through history was not limited to trade goods. Together with the maritime trading routes along the southern coastlines of Eurasia, this extensive overland transport network also provided highways for the diffusion of ideas, philosophies and religions. Breakthroughs in mathematics, medicine, astronomy and map-making, as well as innovations and new technologies including the stirrup, paper-making, printing and gunpowder, all spread between peoples across Eurasia along these trade routes.[31] The integrated overland and maritime networks were the internet of their time, allowing not just commerce over long distances but also the exchange of human knowledge.*

From the sixteenth century, however, the Silk Road began to lose its importance, as trade by land was outcompeted by the global oceanic network knitted together by European mariners in the age of exploration. The ancient Silk Road entrepôts, once some of the most vibrant places on Earth, lost their bustle and shine, and although some of the caravan stops, such as Samarkand and Herat, remain populous cities today, many other trading posts live on only in our cultural memory. It was the coastal ports that began to dominate global trade.

Nonetheless, for centuries the Silk Road was enormously influential for the movement of goods, people and ideas, as caravans threaded their way through mountain passes and across deserts. And it was the ecological zones and landscape of central Eurasia that created a fundamental distinction in the organisation of societies across the continent, leaving an indelible imprint on its history.

---

* This was something of which the more isolated civilisations of the Americas were deprived. When contact was re-established between the peoples of Eurasia and the Americas at the end of the fifteenth century – for the first time since the Bering land bridge was severed at the end of the last ice age – Eurasian civilisation was much more advanced in terms of scientific understanding and technological capability. A shared heritage across the millennia, facilitated by land and maritime trade routes, was one of the major reasons for this faster development.

SEAS OF GRASS

We have seen how the bands of deserts running around the world are created by the dry, descending air of the circulation pattern in the Earth's atmosphere (as well as the rain-shadow effect behind mountain ranges like the Himalayas). But the temperature gradient from the planet's poles towards the equator also defines a series of layered climate zones, and the distinct ecosystems that are found within them.[32] These horizontal stripes on our planet exist in both hemispheres but they are more pronounced in the Northern with its far greater land mass.

Here the northernmost zone, nearest the pole and stretching across northern Siberia, Canada and Alaska, is tundra. Very cold temperatures and a short growing season produce a bleak landscape in which little more survives than scattered dwarf shrubs, heath and hardy lichen clinging to rocks. It is populated only by reindeer herders or caribou hunters.[33]

South of the tundra lies the taiga, a stripe of dense conifer forests. This subarctic ecological zone covers most of Canada, Scandinavia, Finland and Russia, and gradually turns into the deciduous forests of Northern Europe and the United States at its southern limit. Whilst unsuited for agriculture or raising livestock, the taiga has been an important source of furs, including mink, sable, ermine and fox. In early modern history the demand for furs drove trappers across this taiga belt, and it turned Moscow into a major trade hub. In pursuit of furs, Russia expanded eastwards through the fifteenth and sixteenth centuries, across Siberia as far as the Pacific coast and the northern frontiers of the Chinese Qing Empire.[34] In the seventeenth century, French and other European trappers similarly pushed through the Canadian forests.[35]*

---

* Over this period, the Northern Hemisphere experienced a widespread cooling known as the Little Ice Age, and warming furs were therefore highly sought after. We retain vestiges of this chilly period today in the fur-trimmed formal wear of judges and lord mayors, as well as academic gowns, all of which were designed at that time.[36]

South of the tundra, our planet's climate becomes temperate, before turning tropical closer to the equator. This pattern of ecological zones, arranged like stripes between pole and equator, has defined the lifestyles and economic possibilities of the people living within them throughout history. One ecological region in particular has had an enduring influence on the civilisations around the margins of the Eurasian interior.

Sandwiched between the northern, cold climatic band of the taiga and the chains of deserts to the south is a vast tract of grasslands. In Eurasia, this ecological zone is known as the steppes, and they sit in the same band of latitudes as the prairies of North America, with the Argentinian pampas and the South African veld appearing in the corresponding belt in the Southern Hemisphere.

Running through the centre of the Eurasian continental landmass, the steppes aren't affected by the moist oceanic winds, and so receive little rainfall. This makes them too arid for most trees to survive, and the prevailing vegetation is drought-resistant grass. The grass in turn supports great numbers of ungulate mammals (many of which, as we saw in Chapter 3, originally evolved in this ecosystem). The steppes extend more than 6,000 kilometres in a continuous, broad belt from Manchuria to Eastern Europe. They are a vast sea of grass, larger than the entire continental US, but pinched in places into narrow corridors by mountain ranges; as a result they can be roughly divided into three main areas.

The Western or Pontic–Caspian Steppe runs from the Carpathian Mountains and the mouth of the Danube river, bordered by the Black Sea and the Caucasus to the south, all the way to where the Ural Mountains encroach within a few hundred kilometres of the Caspian and Aral seas. (The Great Hungarian Plain forms an island of grassland in the west, cut off from the main belt by the Carpathian Mountains.) The Central or Kazakh Steppe reaches from the Urals to the Tian Shan and Altai mountains, with the Dzungarian Gate in between, through which the northern route of the Silk Roads passed. The Eastern Steppe stretches from Dzungaria, through Mongolia, and along the northern margin of the Gobi Desert into Manchuria, until it reaches the forests that line the Pacific coast.*

The steppes are not an environment well suited to human habitation. Temperatures vary greatly between the seasons. In the dry heat of summer they can rise to 40 °C, and what rain there is falls in heavy thunderstorms. In winter, under the cloudless skies, the steppes become bitterly cold with temperatures dropping to –20 °C or less, the ground gets smothered in deep snow, and howling winds scour the flat landscape. But most significantly, with little vegetation beyond tough grass that our gut cannot digest, the steppes have not much to offer hunter-gatherers and present a formidable barrier to those travelling on foot. To survive in the steppes, you need mobility as well as a way of generating food.

While the camel was ideally suited to Earth's desert band, the grassy steppes stretching across central Eurasia provide a perfect habitat for the horse. The natural range of horses around the world shrank dramatically between 10,000 and 14,000 years ago with the closing of the last ice age. The horse fell extinct in North America, and as the world warmed the species also disappeared from the Middle East. With the retreat of the ice sheets wide areas of dry grasslands across northern Eurasia were replaced by dense forests, and in Europe the horse survived only in a few isolated pockets of natural pasture. But in the steppes of Central Asia, horses and their equid relatives became the most common grazing animals, and here they were hunted

* Today, the sparsely populated Kazakh Steppe offers the perfect location for Russia to launch its rockets from the Baikonur Cosmodrome, the capsules containing the returning crew descending by parachute onto the empty flat plains of this sea of grass. By comparison, NASA launches its missions eastwards across the Atlantic Ocean, and before the Space Shuttle, its capsules would splash down into the North Atlantic or Pacific oceans for their crew to be recovered by ship.

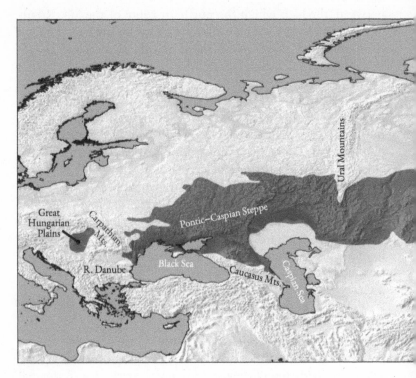

The ecological band of the steppes reaching across the spine of Eurasia.

by the Neolithic tribes. Archaeological evidence indicates that more than 40 per cent of the meat in their diet came from equids.

In fact, horses were at first domesticated not for transportation, but for food. Cattle will not graze grass if they cannot see it through the snow, and the tender nose of sheep allows them to feed only through soft snow. Both are therefore liable to simply stand and starve to death in a winter pasture with forage just beneath their feet. The horse, however, is well-adapted to the cold grasslands and can break even through icy, compacted snow with its hooves to uncover the winter grass beneath; it is also instinctively able to crack through

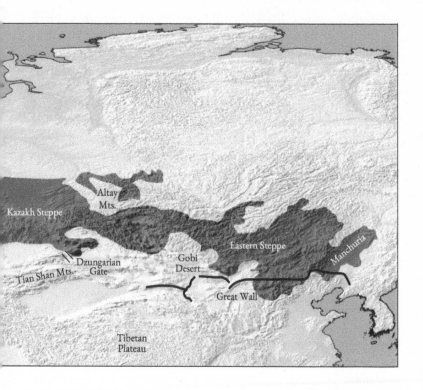

frozen-over water to find drink. Indeed, the trigger for humans to begin domesticating horses may well have been climate change that brought colder winters to Eurasia.[37] This was achieved possibly as early as 4800 BC in the steppes north of the Black and Caspian seas.[38]

Humans learned how to control and ride the animals, and this proved utterly transformative. As we saw in Chapter 3, it was the domestication of grazing mammal species like sheep and cattle that bestowed on humans the ability to convert grass into nutritious meat and milk. Yet settled farmers have limited pasture available for their livestock and this can get exhausted quickly by grazing animals. Herders on horseback with vast grasslands at their disposal could range far further and control much larger herds. In addition, the introduction of solid-wheeled wagons, pulled by oxen, from Mesopotamia[39] around 3300 BC gave the steppe peoples the opportunity to take with them all they needed – food and water supplies, tents for shelter – and roam freely for long periods with their herds across the vast grasslands.[40] This combo package of herbivorous ungulate livestock, fast-moving horseback riding and oxen-pulled wagons serving as mobile homes opened up the steppes for widespread human habitation.*

Irrigated agriculture for cultivating grain is possible only in the more fertile areas along the few rivers that cross the steppes. So on the whole people survived here as pastoral nomads: by raising livestock and constantly moving with their herds between different grazing grounds, following the seasons.[43] And the landscape of the steppes presents few obstructions to overland movements. This core region of Asia is tectonically ancient terrain, uncrumpled by recent plate collisions and rubbed flat by erosion. While the southern margin of Eurasia is characterised by great mountain ranges, the band of the steppes running through the middle of the continent is largely free of such barriers. The exception are the Ural Mountains, one of the rare chains in Asia that run

---

* An important combination of horse rearing and wheeled transport was the construction of the light-spoked and fast-moving war chariot by around 2000 BC.[41] Pulled by a team of well-trained horses and carrying a javelin-thrower or archer, it was the blitzkrieg tank of the Bronze Age. Chariots revolutionised warfare and were as transformational in the conflicts between city states and empires as the later development of gunpowder. But by the time Homer composed the *Iliad* around 800 BC, some five centuries after the Trojan War, this Bronze Age military technology was long obsolete – it had been superseded by tight formations of spear-wielding infantry or fast-moving cavalry armed with the compound bow.[42] The chariot survived merely as a symbol of prestige and power: in Persian, Indian, Graeco-Roman and Norse mythology the gods all ride chariots. Even today many cities have monuments that bear a quadriga, such as the Arc de Triomphe du Carrousel and the Brandenburg Gate.

in a north-south direction,[44] separating the Western and Kazakh steppes and constricting travel to a narrow passage between their bottom tip and the Caspian Sea.* But apart from the Urals, there are few natural barriers like marshes or forests. Horsemen and carts can rove easily across the steppes, turning them into a vast natural highway that sprawls right across the continent and which came to shape the history of Eurasia as a whole.

These nomads entered into an uneasy relationship with the settled agrarian societies around the margins of the continent, ranging from peaceful yet tense coexistence to armed conflict. They traded their herds and their animals' products – cattle, wool from their sheep, and especially the horses they bred in large numbers on the grasslands. They also hired themselves out as mercenaries in the armies of the Eurasian civilisations, often helping to guard their frontiers against incursions from other nomad tribes. They demanded protection money from merchant caravans passing through their lands, or they would ambush them. But they exerted their greatest influence on the course of Eurasian history when they emerged from the depths of the steppes in large numbers to invade the territories of the civilisations settled around the rim of the continent.

Horse-riding nomads were a formidable foe for these agricultural and maritime societies.[46] At times they demanded tributes and could be paid off; at others they raided farms and villages for plunder, and after pillaging what they could carry, they would simply melt back into the vast expanse of the grasslands. Without substantial numbers of their own

---

* The Urals are one of the world's oldest surviving mountain ranges, formed around 250–300 million years ago when the Siberian Plate docked onto the eastern side of Pangea, marking the last stage in the making of the supercontinent. Like the pristinely preserved ophiolite scooped out on top of Cyprus that we discussed in Chapter 6, the Ural mountain range also contains flecks of crust from a long-since vanished ocean and so holds a legacy of rich copper mines.[45]

mounted cavalry, the armies of the agrarian societies could not pursue them into the sea of grass, the arid plains offering no food to support campaigns of foot soldiers. And repeatedly throughout history, loosely aligned in large confederations, nomadic tribes erupted out of the steppes to invade and conquer the settled civilisations, at times forging huge empires that stretched across Asia.[47]

Yet the influence of the steppe peoples on the civilisations of the Eurasian rim did not just consist of direct military assault. As nomadic herders, they were constantly on the move; but when there were disruptions to the delicate equilibrium of their environment – a surge in population numbers or a deterioration of the grazing grounds due to a shift in climate – entire tribes were forced to migrate from their existing grounds to seek better pasture. As a result waves of disruption rippled across the steppes when a succession of displaced tribes moved across the flat plains pushing their neighbours off their turf, like billiard balls recoiling off each other. Eventually some steppe peoples would be forced to cross into the lands of the settled societies, for example Manchuria and northern China in the east, Ukraine and Hungary in the west.[48]

Thus the history and fate of civilisations around the margins of the great Eurasian landmass – in China, India, the Middle East and Europe – has been the story of a recurring struggle against nomadic tribes emerging from the central heartland of the steppes. The Scythians were among the first to master mounted warfare. They originated around the Altai Mountains and came to dominate much of the steppe region between the sixth and first centuries BC, riding westwards to confront the Assyrian and Achaemenid empires in Mesopotamia and Persia, and also fighting Alexander the Great. China was repeatedly confronted by steppe peoples, including the Xiongnu, Khitans, Uighurs, Kirghiz and Mongols.[49] And between the fifth and sixteenth century AD a succession of nomad groups poured from the steppes into Europe – the

Huns, Avars, Bulgars, Magyars, Kalmuks, Cumans and Patzinaks, as well as the Mongols.[50]

For millennia, the steppes represented a great churning cauldron of pastoral nomads, repeatedly brimming over its lip to spill into the domains of the settled, agrarian civilisations around the continental margin. This conflict between the two was an enduring dynamic of Eurasian history, and is fundamentally born of an ecological distinction between dry grasslands and fertile agricultural lands – the worlds of the steppe and the sown – and the different human lifestyles they support. But it was the landscape of the continent that shaped and channelled these migrations and invasions along the same courses time after time.

DISPLACED PEOPLES

In the same way that the Silk Road passed through narrow corridors, valleys and mountain passes, the landscape provided convenient passageways for armed raiders to cross into the lands of civilisation. If these channels facilitated trade along the overland routes, they also made the settled societies around the Eurasian rim vulnerable to raids and conquest.

India was largely protected by the great barrier of the Himalayas, but the narrow Khyber Pass through the Hindu Kush provided an entry point for invaders. China, as we saw earlier, also generally benefited from natural barriers, but its central plains were open to nomad incursions from the steppes to the north, and from the west through the Dzungarian Gate, which leads invaders along the Gansu Corridor into the heartlands of China.[51]

The Great Wall was built to defend China against the influx of nomads from the steppes. After unification, the first emperor of the Qin dynasty fortified this northern frontier from 221 BC,[52] and the walls were extended by the Han between 200 BC and AD 200 to guard the stretch of the Silk Road that passed along the Gansu Corridor to the Tarim Basin. However, much of the most impressive remnants of the wall date to the construction during the Ming dynasty from the mid fourteenth century. On the face of it, the Great Wall serves as a demarcation line between two elementally different lifestyles and cultures – the nomadic and the settled, the barbarians and the civilised. But in a deeper sense, these fortifications were built along the fundamental ecological boundary between the wet, fertile lands supportive of agriculture and the dry, harsh steppes in the heart of the continent, where only pastoralists could survive. Nonetheless, China was repeatedly invaded by steppe peoples, often entering through the Dzungarian mountain pass and along the Gansu Corridor. Just as the Khyber Pass provided a point of entry into India for nomadic raiders, China was also attacked along the route of the Silk Road. The passageways of trade also facilitated invasion.

On Eurasia's western edge, Europe is vulnerable to incursions and invasions along a few major low-lying routes and highland passes that provide access to nomads from the steppes. From the western steppes, one route passes south of the Caucasus and Black Sea along Anatolia; another heads north of the Black Sea to the Carpathian Mountains, and then either north between these mountains and the Pripet Marshes, or south following the Danube Valley, both ways taking invaders into the heart of the Northern European plains.[53] The Huns that challenged the Roman Empire from the fourth century AD, the Bulgars migrating into the Balkans in the seventh century, the Magyars entering the Hungarian plains in the ninth, and the Mongol invasion of the thirteenth century, all originally approached Europe from the steppes along these corridors.[54]

If the clashes between nomadic tribes and settled societies reflected the lifestyle supported by their respective habitats, the natural world and distribution of the different ecosystems also dictated the steppe nomads' course of action after they had invaded the agrarian lands.

The formidable threat presented by the horse peoples was largely attributable to their mobility. Unlike the slow-moving armies of the settled civilisations, nomads were able to operate swiftly over vast distances. But the steppe raiders were bound by a fundamental ecological constraint. Their military prowess depended on being able to field great numbers of fast-moving mounted warriors, but their horses needed to be fed. This was simple in their natural habitat of the vast grasslands of the steppes, but as soon as the invaders penetrated too far into the agricultural lands around the Eurasian rim they struggled to nourish their mounts. Irrigated farmland yields plenty of grain from a small area for feeding people, but it is not effective as pasture for supporting great numbers of horses.

This constraint imposed by Nature reveals that the arable and pastoral lifestyles are intrinsically incompatible, and so after enjoying the spoils of victory the invaders from the steppes were either forced to withdraw back to their expansive natural pastures, or fundamentally change their ways and assimilate into the settled society.[55] It should not surprise us, therefore, that the Huns, who invaded the very heart of Europe in the mid fifth century AD, chose as their centre of operations the Hungarian Plains – an ecological borderland between steppe and agricultural fields, and the westernmost pocket of the steppe grasslands.[56]

Others abandoned their nomad ways. The Ottoman Turks were originally pushed out of the steppes and into Anatolia in the thirteenth century by the Mongol expansion of Genghis Khan. Here they established themselves by adopting a European style of warfare relying on fortifications, and formed slave armies of captured Christian boys who were forced to convert to Islam, the famous Janissaries.[57] By the end of the thirteenth century the Ottomans had become a major threat to the states of Christendom, and in 1453 they captured Constantinople and brought to an end the Byzantine Empire.[58]

Nomadic horsemen riding out of the steppes were the cause of two of the most defining moments of world history: the fall of the Western Roman Empire, and the conquest of Asia by the Mongols.

DECLINE AND FALL OF THE ROMAN EMPIRE

We saw earlier how by the first century AD the Roman Empire had expanded around the rim of the Mediterranean, stopping at the natural borders defined by the deserts of North Africa and the mountain ranges and large rivers in Europe. By AD 300, however, the full length of the empire's north-eastern border along the Rhine and Danube rivers was under pressure from the growing population of Germanic tribes occupying the wilds beyond. The situation worsened a few decades later with a series of violent incursions and forced migrations triggered by a horse people emerging from the steppes that pushed these tribes over Rome's borders. These horse people are widely believed to be the same confederation of nomadic tribes which on the eastern limit of the steppes belt had been challenging China since the third century BC:[59] the Xiongnu. When they appeared in the west, they came to be known as the Huns.[60]

The Huns now moved westwards across the steppe belt, in all likelihood seeking better pasture during a period of regional climate change – we have evidence of a cooling in the Northern Hemisphere at that time that caused droughts in the steppes[61] and would have diminished grass resources for feeding their flocks and horses. The Huns reached the Don river by the 370s AD,[62] in the process displacing other nomad groups who in turn drove settled villagers off their lands in Eastern Europe.

Huge numbers of these refugees arrived at the frontier of the Western Roman Empire along the Rhine and Danube rivers, and before long tribe after tribe began to pour into Roman territory – Burgundians, Lombards, Franks, Visigoths, Ostrogoths, Vandals and Alans.[63]

By the end of the fourth century, after driving a succession of tribes in front of them like a great bow wave of peoples fleeing out of the way, the Huns themselves arrived at the borderlands of the Roman Empire. They set about conquering the tribes living north of the Danube before turning on the Eastern Roman Empire, which had been spared much of the earlier tribal migrations and invasions. Led by the fearsome Attila from 434, the Huns ravaged Greece and the Balkans in successive campaigns, and reached the walls of Constantinople itself. They were stopped by the city's formidable fortifications, but nonetheless were able to exact huge tributes from the Empire.

Emboldened by these successes in the east, Attila now turned his aggressions to the Western Roman Empire. Advancing along the Danube and Rhine, sacking city after city along the way, he invaded Roman Gaul (modern France) in 451, before being beaten in battle by an alliance of the same tribes and horse peoples who had been originally displaced by the Huns' emergence from the steppes. But Attila returned the following year to devastate the plains of northern Italy, and forced the emperor to agree terms to stop the Huns from marching on Rome. Attila died two years later and the Hun Empire dissolved shortly afterwards but they had already set in motion the wheels for the destruction of the Western Roman Empire.[64]

And it wasn't just the Romans who felt the brunt of these displaced peoples. Persia too experienced an onslaught of nomadic tribes spilling over the Caucasus and sacking the cities of Mesopotamia and Asia Minor.[65] At the very end of the fourth century the Eastern Roman Empire and Persia, facing a common enemy, put aside their long-running enmity to work together in the construction and garrisoning of a huge fortified wall. Running for around 200 kilometres from the Black Sea to the Caspian Sea, the wall was fronted by a

4.5-metre-deep ditch and studded with 30 forts along its length, manned by 30,000 troops. This Persian wall is second only to the Great Wall of China as the longest defensive barrier ever constructed, and was built for exactly the same purpose: to defend the boundary between settled civilisation and the barbarian wilderness.[66]

But for the Western Roman Empire it was already too late. The frontier along the Rhine and Danube had become overwhelmed, and wave after wave of migrating tribes crashed through the defences. The Visigoths marched down through the Italian peninsula and in 410 sacked the city of Rome itself.[67] The Vandals, another tribe that had been displaced by the Huns, advanced though Central Europe, crossed the Iberian Peninsula and invaded Roman North Africa to seize in 439 the city of Carthage and the surrounding regions that had supplied grain to the Western Empire. Their conquests also included Sicily, Sardinia and Corsica, and in 455 the Vandals too sacked Rome. By 476, the centralised control of the Western Roman Empire had effectively dissolved, its former territories now divided into kingdoms ruled by the Germanic tribes that had flooded across the imperial frontiers from the east – the Franks in France and Germany, the Visigoths in Spain, and the Ostrogoths in Italy. Through the Middle Ages these kingdoms developed into the nations of modern Europe.

The Western Roman Empire had been destroyed by the 'great migration' of settled tribes and pastoralists from the steppes. Once again, it is fundamental planetary causes that explain this turning point in history. Ultimately, the fall of Rome was due to the ecological distinction between the arid grasslands of the Eurasian steppes that sustain horse-riding pastoral nomads and the wetter lands around the rim that supported the settled agriculture of the empire, and a climate shift within the steppes that triggered these waves of displaced peoples.

PAX MONGOLICA

In the thirteenth century, horse people from the steppes again changed the course of history across Eurasia. The Mongols emerged from the grasslands and in just twenty-five years they succeeded in conquering more territory than Rome had annexed in four centuries.[68] The Mongol Empire not only united the tribes of the vast Eurasian steppe, but also included China, Russia and much of South West Asia, making it the largest land empire the world has ever known.[69] The leader who instigated this spectacular campaign was the son of a prominent tribal chief in eastern Mongolia, born with the name Temüjin (perhaps meaning blacksmith). But it is his adopted title by which he has become (in)famous: the 'fierce ruler' – Genghis Khan.[70]

Genghis Khan belonged to just one of many nomadic tribes herding sheep on the northern fringe of China, but by 1206 he had unified the surrounding tribes and become master of the Mongolian steppes.[71] With his power base consolidated, his hordes of mounted raiders now thundered out of the steppes to attack the civilisations around the rim of Eurasia. They invaded northern China in 1211[72] and then swept through Central Asia.[73] Genghis Khan died in 1227, but his successors proved just as successful in their military expansion.[74] The Mongol conquest continued across the Middle East before the tribesmen headed up through the Caucasus to southern Russia and into Eastern Europe.[75]

Here they advanced into Poland and the plains of Hungary, reaching the outskirts of Vienna[76] and spreading panic throughout Christendom.[77] But Europe was spared by a fateful twist of history. The Great Khan at that time, Genghis' son and successor Ogodei, died suddenly, and the Mongol leaders withdrew to their capital in Karakorum to select their next supreme ruler. In the end, the khans did not attempt to continue their conquest towards the Atlantic – the Mongolian empire effectively ended at the western terminus of the steppe belt.[78] Instead they turned east again, conquering China in its entirety and establishing themselves as the Yuan dynasty.[79] Its first emperor, Kublai Khan, ruled from Shangdu – spelled as Xanadu by Coleridge in his famous poem – before moving his throne to Beijing.[80]*

By the end of the thirteenth century the Mongol Empire stretched across the entire breadth of Asia, from the Pacific Ocean to the Black Sea. During this incredible expansion, the Mongols had been notoriously savage in their treatment of cities that refused to surrender immediately.[82] They slaughtered all inhabitants – men, women, children, as well as livestock – leaving behind only empty streets and pyramids of skulls. This deliberately gruesome wielding of violence was designed to encourage the next cities in their path to capitulate without a struggle – the horrifying reports of their savagery spread ahead faster than the advance of their armies. But the Mongols weren't just the fearsome hordes of ferocious warriors of the popular conception. Once resistance had been subdued, captured towns and cities were often rebuilt under the Mongols' careful stewardship.[83] The khans were also remarkably tolerant of the different peoples they ruled over, permitting cultural and religious freedom.[84] After the initial campaign of shock and awe, the Mongols were able to win over hearts and minds.

Moreover, when the initial fury and violence of conquest had passed, the unification of Asia produced an era of booming trade across the breadth of the continent. This has come to be known as the 'Pax Mongolica' – echoing the Pax Romana, the period of stability and prosperity around the Mediterranean

---

* On numerous occasions throughout the thirteenth and fourteenth centuries, the Mongols also invaded the north-western region of India, but it wasn't until 1526 that one of Genghis' descendants established the Mogul Empire across the subcontinent.[81]

during the Roman Empire a millennium earlier. For around a century from 1260, the Mongol khanates ensured the secure passage of merchants across Asia, and their skills in administration and savviness in keeping taxes low all combined to foster commerce.[85] In contrast to the smash-and-grab tactics of earlier nomad invaders who relied on plundered loot or tributes extorted from the agrarian civilisations, the khans appreciated that they could profit far more from trading than from raiding. Commerce along the Silk Road flourished during this period, the caravans not only striding along the old desert routes of Central Asia but also turning further north to the Mongolian capital Karakorum and across the grassy steppes.[86] The Mongols had accomplished the linking of east and west like no one before.

As a result, spices and other luxury goods poured into Europe.[87] The blast furnace arrived in the west during the Pax Mongolica, and the Mongols also introduced Chinese gunpowder to Europeans,[88] changing for ever the nature of warfare. But the unification of Asia and the ease of movement across the continent had another profound ramification for history. Something else far more destructive also entered the bloodstream flowing along the communication arteries across Eurasia: disease.

The Black Death emerged from the steppes and surged across this connected world in the mid fourteenth century. Bubonic plague reached China by 1345 and Constantinople in 1347. From there it travelled to Genoa and Venice aboard merchant ships,[89] and had spread to Northern Europe by the following summer. People already weakened by poor nutrition due to a series of harvest failures – the arrival of the plague coincided with the beginning of the first cold spike of the Little Ice Age[90] – quickly succumbed to the disease. Within just five years the Black Death had killed at least a third of the European and Chinese populations,[91] and also devastated the Middle East and North Africa. Around 25 million people died in Europe alone.[92]

The plague hit the Mongol khanates just as hard, their grip on power already weakened by internal rivalries. In China, the Yuan dynasty was overthrown by the Ming in 1368, and across Eurasia the vast Mongol Empire again splintered into many states without political or economic unity. The steppes became once more a mosaic of jostling nomadic tribes, and the highway between East and West crumbled. But in Western Europe, the aftermath of the Black Death brought some beneficial outcomes. The severe depopulation meant that many lords had lost the tenants on their land, and so were forced to accept lower rents and a more mobile peasant workforce. The shortage of labour also meant that craftsmen and agricultural workers could demand higher wages. This relaxed the serfdom under the feudal system and improved social mobility in Western Europe where the guilds in the more populous, mercantile towns already had considerable sway.* The disruption of the Black Death, which had emerged from the steppes and spread with the help of the trade infrastructure maintained by the Mongols, shook the foundations of feudalism and helped create the beginnings of a different and more mobile society.[93]

And the conquests of the Mongolian superpower had other far-reaching consequences for the history of Europe. As they had surged westwards they had destroyed the great Islamic empire of the Khwarezmids in Central Asia, massacring their trading entrepôts of Samarkand, Merv and Bukhara, as well as devastating Baghdad, the capital of the Abbasid Caliphate. But crucially, the Mongols had stopped short of advancing far through Europe. The ports of Venice and Genoa remained the major mercantile centres of the West and grew in wealth and power through the later medieval period and Renaissance. By destroying the old Muslim core of Eurasia but sparing Europe, the Mongols had tipped the power balance in the region and Europe was given the opportunity to pull ahead and begin to

---

* In contrast, the landowners in Eastern Europe held more power and were able to force the remaining peasantry into tighter serfdom.

develop faster than the Islamic world.[94] Still, when Constantinople fell to the Ottomans in 1453, the Byzantine Empire had been little more than a rump state for more than a century, with Muslim rulers dominating the whole eastern Mediterranean and blocking the trade routes from the East into Europe.[95] It was for this reason that European mariners began to look west for new maritime routes to the riches of China and India in the Age of Exploration, as we'll see in the next chapter.

END OF AN ERA

For millennia, the steppes had represented a vast wilderness, home to nomadic pastoralists. These grasslands had supported large numbers of horse-riding warriors, able to strike at the agricultural civilisations around the rim of Eurasia in devastating raids. But from the mid sixteenth century, first the states of Renaissance Europe, and then Russia and China, began to decisively shift the balance of power between the worlds of the sown and the steppe. The pivotal development was that of a whole system of interrelated advances known as the Military Revolution. The agrarian states learned to effectively use gunpowder for musket and cannon, developed coordinated military drills to deliver devastating firepower on the battlefield, established far-reaching logistics to keep their troops supplied, and transformed their economies to support larger and larger standing armies.[96] These innovations centralised military power, allowing rulers to consolidate their control and unify fiefdoms into large single states, marking the beginnings of our modern nations.[97]

Steppe societies were unable to compete with this military progress. While they could trade for firearms, just as throughout history agricultural societies had bought horses from the steppes, their purchasing power was restricted by a far less developed economy than that enjoyed by the consolidated agrarian states. This tipped the balance for the first time squarely away from the pastoral nomads in favour of the settled societies. The final gasp of nomad power was drawn in the 1750s with the defeat by Qing China of a confederation of Mongol tribes in Dzungaria. The military threat from the steppes had finally been contained and a long chapter of Eurasian history drew to a close.[98] Never again would an empire of nomads emerge from the steppes and trigger an existential crisis among agrarian civilisations.

On the contrary, it was agrarian civilisations on the margins of the steppes which now began to penetrate further and further into these open grasslands, settling them and cultivating the soil, and thereby further strengthening their economies.[99] Russia and China expanded into this middle ground, until their borders came to abut each other.[100] Russia in particular grew into a great superpower by expanding into the steppes formerly ruled by the Mongol Empire, not in search of pasture for livestock and horses, but to exploit the rich mineral resources of this huge region and to turn it into high-yielding farmland, making use of a fertile loess soil that had been further enriched with nutrients by the grasses which had grown here for millennia.[101] The expanding Russian Empire gradually transformed the Pontic–Caspian Steppe north of the Black and Caspian seas into vast fields of golden, swaying wheat.[102] And by the 1930s these lands had become of immense strategic importance.*

---

* We've focused in this chapter on the steppes running across the spine of Eurasia, but the same ecological region also appears in North America. The prairies reach in a broad stripe right down the centre of the United States, occupying the dry region of the continental interior and rain shadow of the Rockies. As we've already explored, compared to Eurasia, North America was impoverished in its biological heritage. The horse had fallen extinct in its birthplace and there were no cows or sheep that supported the steppe nomads. The dominant mammal of the prairies, the bison, was hunted by indigenous American tribes, but resisted domestication. Plants like the squash, as well as several seed-bearing species such as the sunflower, had been

The major motivation for Hitler's invasion of the Soviet Union in June 1941 was not just to seize the vital oilfields of the Caucasus region but also to lay claim to the fertile farmland of the former steppes to the north. These were both to provide their huge agricultural potential and satisfy Hitler's vision for securing *Lebensraum* – 'living space' for the continued survival of the German people.[105]

Operation Barbarossa ultimately failed, with the Wehrmacht defeated as much by the challenges of logistics across such vast distances as by the onset of the bitterly cold winter of the steppes and by the Red Army. But Hitler's ambitions illustrate powerfully how the terrain of the steppes has been profoundly transformed over the past few centuries – from a wilderness inhabited by horse-riding pastoralists threatening the settled civilisations of Eurasia to rich, cultivated farmland now vital for feeding these same agricultural societies.*

The long era of Eurasian history with nomadic societies from the steppes repeatedly clashing with the civilisations around its rim was born of an ecological and climatic distinction, with contrasting regions supporting either horseback pastoralism or settled agriculture. The overland trade routes across the deserts of North Africa and Arabia, and the Silk Road linking the breadth of Eurasia, were also dominated by a particular climate zone – that of the band of deserts created by the dry, descending

---

domesticated by about 4,000 years ago in eastern North America,[103] but the prairie belt represented a huge area of the Earth that was never utilised for agriculture.[104] This all changed after the European conquest, with the arrival of colonists who brought with them the livestock and crops domesticated in the Old World. The drier prairies in the west proved perfect for open ranching of cattle, and over the last two centuries, with the help of steel-edged ploughs, advanced irrigation techniques and artificial fertilisers and pesticides, the eastern prairies have become some of the most productive croplands on the planet.

* In 2016, Russia became the world's biggest wheat exporter, much of the harvest coming from the steppe region north of the Black Sea, and supplied to the Middle East and North Africa.[106]

arm of one of the great circulation patterns in the Earth's atmosphere. The global circulation patterns are also responsible for the prevailing winds around the world, and these the Europeans charted and learned to exploit during the Age of Exploration to create huge oceanic trading networks and powerful overseas empires.

Chapter 8

The Global Wind Machine and the Age of Discovery

The Age of Exploration began on the Iberian Peninsula, the very western extremity of Eurasia, peripheral to the exchange of goods and knowledge across the continent. The kingdoms that were to become Portugal and Spain could only regard with envy the riches trafficked by ports like Genoa and Venice across the Mediterranean. Through the Middle Ages, much of Iberia had been under Islamic control, after the Umayyad Caliphate had invaded across the Strait of Gibraltar in 711.* The Christian kingdoms of the peninsula pushed back during the centuries of the Reconquista, with Portugal securing the full extent of its kingdom along the western coast by the mid thirteenth century. But it remained hemmed in by its larger and richer neighbour, Castile, and faced only the unknown expanse of the Atlantic.

The Portuguese continued their holy war across the Strait of Gibraltar, and in 1415 captured the Muslim port of Ceuta on the northern tip of Morocco, one of the end-points of the trans-Saharan

* The modern name for Gibraltar is derived from the Arabic *Jabal Tariq*, Mountain of Tariq, after the Islamic general who led this invasion. In the ancient world, Gibraltar formed one of the two Pillars of Hercules – the other being Mount Abila on the North African coast – that marked the beginning of the end of the known world. With European expansion into the Atlantic the Strait of Gibraltar became a vital naval chokepoint, controlling the passage into the Mediterranean Sea.

caravan routes. It was here that the Portugese first tasted the wealth that might be acquired if they could outflank the Muslim world and carry this gold and slave trade on their own ships.[1] They began to explore the West African coastline to find the sources of the gold, and before long some mariners contemplated the possibility of sailing all the way round the southern tip of Africa to reach India and the riches of the spice trade.[2]

Then, by the late fifteenth century, the kingdoms of Castile and Aragon united into what would become modern Spain. In 1492 they completed the Reconquista of the peninsula when they captured the last Moorish stronghold of Granada, and joined Portugal in seeking new overseas trade routes and territories through the Atlantic.*

VOLTA DO MAR

In the Atlantic, at some distance from the European and African coasts, lie four small archipelagoes: the Canary Islands, the Azores, Madeira, and the Cape Verde Islands. For the Romans the Canary Islands marked the end of the known

---

* The reason why Spain joined the Age of Exploration significantly later than Portugal also comes down to plate tectonics. As we've seen, the Mediterranean is a tectonically complex region, formed by the disappearance of the Tethys Sea as Africa rammed north into Eurasia, and with a jumbled mess of small fragments of continental crust becoming caught in the collision zone. One of these is the Alborán microcontinent, which over the last 20 million years has moved west to crunch into the south-eastern margin of Spain and thrust up the Sierra Nevada mountain range.[3] It was here, in this easily defensible rugged terrain, that the last bastion of Islamic rule, the Emirate of Granada, held out for another 250 years after the rest of the Iberian Peninsula had been reclaimed by the Christian Reconquista. While the Kingdom of Portugal, occupying the flatter terrain across the western side of the peninsula, had secured its territory by the middle of the thirteenth century, and was able to invest its energy in maritime exploration, Spain remained preoccupied with its own, more challenging reconquests until the very end of the fifteenth century.

world,* but knowledge of them seems to have been lost during the Dark Ages: they literally vanished off the maps. They were rediscovered and encountered alongside the other, previously unknown archipelagoes, in the late fourteenth and early fifteenth centuries when Portuguese and Spanish sailors began venturing beyond the Iberian Peninsula.[4] They found the Canaries, only about 100 kilometres off the coast of Morocco, to be already settled by indigenous tribes, probably descendants of Berbers from North Africa, but the more remote Azores and Cape Verde Islands were uninhabited when the Portuguese reached them.

Iberian sailors heading out to sea soon encountered the Canary current that carried them south-west down the African coast. At about 30° latitude, prevailing northeasterly winds picked up to carry them to the Canary Islands. This course along the Moroccan coastline, borne by favourable currents and winds, was an ancient sea route, used by the Phoenicians to trade along the north-western coast of Africa with their galleys, which were also fitted with banks of oars. The problem for European sailors venturing out 2,000 years later was how to get home again. Sailing ships don't need teams of straining oarsmen and can therefore carry more provisions and trade goods, but they struggle to make headway against adverse currents or winds.

The critical innovation developed by Portuguese navigators is known as the *volta do mar* – the turn, or return, of the sea. In order to get back north-east to Portugal from the Moroccan coastline or the Canaries, they turned out west into the expanse of the Atlantic Ocean. This seems pretty paradoxical at first but the Canary current weakens further offshore, and as soon as the ships were north of about 30° latitude they could pick

---

* Their name derives from the Latin 'Islands of the Dogs', although this description may have actually referred to the large seals that once packed the archipelago's beaches. Canary birds in turn were named for these islands to which they are indigenous.

up prevailing southwesterly winds and ride them all the way back home. Thus, on their return voyage to the Canaries these sailors took advantage of different regions of oceanic currents and atmospheric wind circulation. It just so happens that the Canary Islands lie close to the region on Earth where the northeasterly trade winds give way to the southwesterlies.

We'll come back to this later, but at this point it's worth explaining a perplexing quirk in the way winds and ocean currents are named. A wind is specified by the direction from which it is blowing, so a northerly wind blows from the north towards the south. Ocean currents, on the other hand, are named the opposite way: by the direction in which they are going. Thus a northerly current arrives from the south and carries you north. This is potentially very confusing, but it does carry a degree of sense. When you're on land, the direction a wind is coming from is the important aspect: what matters is from where a storm arrives, or the direction in which you need to turn a windmill. But for a ship being carried along by an ocean current, it's where it's taking you that is important – especially if it's towards a reef or shoal that could wreck you.[5]

If you steer a wide, looping *volta do mar* course through the open ocean to return to the Iberian Coast from the Canaries you will get to Madeira. Although Madeira actually lies closer to Portugal, the Canaries were discovered first as the prevailing northeasterly winds carried the European ships straight there from the Strait of Gibraltar. As successive Portuguese expeditions pushed further and further down the African coast, they steered wider *volta do mar* courses out into the mid Atlantic, and in the process encountered the Azores. This archipelago lies about 800 kilometres from the edge of the Iberian peninsula, and from here another ocean stream, the Portugal current, carried the ships back to port. Finally, the Cape Verde islands, lying off the western bulge of the African continent at the point where the Sahara desert yields to the thick tropical rainforest of Central Africa – the islands' name means the 'Green Cape' – were discovered by the Portuguese in 1456.

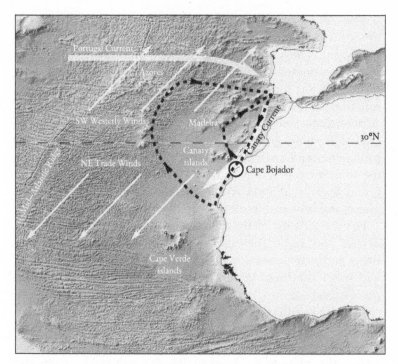

The Atlantic archipelagoes and example *volta do mar* routes exploiting different regions of winds and currents.

Unlike islands such as the Isle of Wight, Mallorca or Sri Lanka, which sit on the continental shelf but have been cut off from the mainland by rising sea levels, these Atlantic archipelagoes stand isolated in the ocean: they are the very tips of volcanoes rising from the sea floor.[6] In fact, the Azores are the peaks of the highest volcanoes of the Mid-Atlantic Ridge, the great spreading rent in the oceanic crust that stretches all the way to Iceland.[7]*

---

* Isolated volcanic islands have played an important role in history, offering strategic value as specks of land within the great expanse of the ocean. Saint Helena in the South Atlantic is another volcanic island born of the Mid-Atlantic Ridge and one of the most remote islands in the world. It

The Atlantic islands provided important oceanic staging posts for the Iberian explorers – they were stepping stones in the ocean. The Canaries in particular presented a vital stop for taking on board provisions and fresh water, enabling ships to embark on longer voyages.[8] The Azores offered a similar function on the route home. The early navigation between the African coast and these different archipelagoes also served as crucial training grounds, allowing European sailors to gain proficiency and confidence for attempting greater voyages into the unknown. It was here that they began to understand the grand-scale circulation within the planet's oceans and atmosphere, and how to exploit these patterns of currents and winds.

But the Atlantic islands also became economically valuable in their own right. Their climates and rich, volcanic soils made them perfectly suited for growing crops like sugar.[9] Madeira had originally been densely forested – it took its name from the Portuguese for 'wood'[10] – but the forests were rapidly cleared by Portuguese sailors and the land turned to the cultivation of wine and sugar. By the end of the fifteenth century, Madeira was producing almost 1,400 tonnes of sugar every year,[11] the plantations worked by slaves brought from the African mainland. So the Atlantic islands played a pivotal role in the Age of Exploration, but their 'discovery' also presaged the ugliest aspects of European expansion: territorial conquest, colonialism and plantations worked by slave labour.

---

became a vital stop-off point for ships of the East India Company returning from India and China, and it was here that the British imprisoned Napoleon after his final defeat at the Battle of Waterloo. In modern history, the volcanic chain of the Hawaiian archipelago in the middle of the Pacific Ocean was of great strategic importance to the United States, which established airfields and naval bases there. When the Japanese attacked the ships stationed at Pearl Harbor in the lagoon of the island of Oahu in December 1941, the incident drew the US into the Second World War. The bombing raids flown from Midway, one of the most north-western islands of the long Hawaiian chain, as well as from American aircraft carriers, crippled the Japanese fleet six months later and proved to be decisive in turning the war in the Pacific.

TO THE CAPE OF STORMS

If you look at a map, Cape Bojador appears as no more than a bump in the convex coastline of West Africa. Yet for a time this innocuous-looking, duney headland was considered the southernmost point to which it was possible to sail along the African coast – treacherous to navigate, it was known in Arabic as Abu Khatar, 'the Father of Danger'.[12]

The maritime tradition of the time was for ships to hug the coast. Staying close to the shoreline gave regular access to food and fresh water and, more importantly, provided landmarks for navigation. But around Cape Bojador the gentle winds blowing along Morocco give way to strong winds from the east that threaten to blow any ship out into the open ocean.* In addition, a broad, submerged sandbar at the cape extends from the shore over 20 miles out to sea, reducing the water depth to just a few metres. Thus a ship leaving sight of the coastline to steer around that danger risks being caught by the stronger currents and dragged further and further out into the ocean.[13]

But then, in 1434 the Portuguese navigator Gil Eanes came up with a revolutionary new technique that allowed him to round Cape Bojador – known today as current sailing. To sail in a desired direction through complex winds and oceanic currents you need to take into account the deflection to your ship's course by the unseen current. The only way Eanes could have done so is to have meticulously measured both the direction and speed of the current at the Canary Islands before setting off, and then at several points along the way by hauling in his sails or dropping anchor to get a reading on the local

---

* Most air currents are invisible, but in this case the winds can be clearly seen from space, thick with dust picked up from the Sahara Desert. The dust-laden air takes about a week to cross the Atlantic, and then the particles settle to fertilise the rich soils of the Amazon rainforest.

current and make the necessary corrections to his course. Eanes may have initially guessed the compensatory course he needed to sail, or perhaps he even calculated it, as modern mariners would, by plotting a triangle on his nautical map: marking the line between his current position and his destination, the line of deflection from the current, and joining these with a third line that showed the actual course that must be steered to compensate for the current. Cape Bojador was thus conquered by Portuguese navigators seeking to understand the patterns of the sea. And as they began to master them, they gained confidence to sail further from the shore.

Once the way past Cape Bojador had been shown, successive Portuguese expeditions pushed steadily further down the coast of West Africa, discovering the Senegal river as well as the Cape Verde archipelago lying 570 kilometres offshore. By 1460 the Portuguese had sailed 3,000 kilometres down the African coast, and were now edging round the lip of the great bulging protrusion of West Africa to enter the Gulf of Guinea.[14] Here the Guinea Current carried them east, but the explorers found that the prevailing northeasterly winds that had been a reliable companion all the way south since the leaving the Canaries gave out. They now had to contend with the light and variable winds of the doldrums.

In 1474 Portuguese captains reached the point where the African coast turned south again, and when soon after they crossed the equator they lost sight of Polaris, the 'pole star'. This is a bright star in the constellation of Ursa Minor (the Little Bear, or Little Dipper) that just happens to lie directly above the North Pole. If you want to work out your latitude – how far north of the equator you are – you simply measure the angle between Polaris in the night sky and the horizon. But as the star now disappeared from view, the sailors were entering not just uncharted waters but strange new parts of the world where even their navigational techniques no longer worked. The Portuguese word coined for losing sight of Polaris was *desnorteado* – to be 'dis-northed' – and it soon took on the more general meaning of being lost or confused.[15]* But as Portuguese sailors continued down the African coast, on the opposite horizon they caught sight of the Southern Cross, a bright constellation that could serve the same guiding function in the Southern Hemisphere.[16]

As the Portuguese continued with their quest to find the southern tip of this mysterious continent, each mission stopped regularly to collect information about the local geography, languages and, crucially, the goods that could be traded. Their ships also took with them stone pillars to be erected at the furthest point along the coast they reached on each expedition. Not only were these intended to stake territorial claims for the glory of the Portuguese crown, but they also served as a visible marker to be exceeded by subsequent voyagers.[17] These small monuments carried in the hold of the pitching and rolling caravels as they sailed towards new frontiers were the fifteenth-century equivalent of the flags carried by US astronauts on the Apollo missions to the Moon.

Yet the first successful rounding of the tip of Africa was a step change from this slow probing along the coast: it required a radical new approach.

In the late summer of 1487, Bartolomeu Dias set out from Lisbon, passed the Canary Islands, rounded Cape Bojador, and retraced the route along the African coastline now familiar from decades of Portuguese exploration. After four months at sea, Dias had passed the stone pillar marking the furthest point previous expeditions had reached. As he continued to follow the shoreline, he named the bays and capes that he encountered after the saints' days: the Gulf of Santa Marta (8 December), São Tomé (21 December), Santa Vitória (23 December), and so on, like time-stamps marking his progress on the map. On

---

* The equivalent English word, derived from the French, for losing your bearings is to become 'disoriented' – to have lost the eastward direction, from where the sun rises.

Christmas Day he christened the Gulf of St Christopher, the patron saint of travellers.[18]

All along this coastline Dias' ships had been pushing against adverse flows, tacking both against a steady southerly wind and a current pushing north up the coastline. Then Dias made a radical decision. He turned his ships away from land and headed out into the vast ocean, watching the comfort and safety of the coast recede over the horizon. His hope was that the same trick necessary for returning home from the North African coast against the Canary current – heading out further to sea on a looping *volta do mar* course to pick up westerly winds – would also work in the South Atlantic to carry them round the southern tip of Africa and find the passageway to the east.

Dias's flash of insight paid off, and by about 38° south the hoped-for westerly winds began to pick up. The ships finally turned east with these winds, and after nearly a month in the featureless expanses of the South Atlantic, they at last made landfall. Following the shoreline they realised that the coast was now heading north-east: they had successfully rounded Africa's southern tip and were on the far side of the vast continent. But with their onboard provisions running out, Dias was forced to erect his final marker pillar and turn back home. It was only on the way back that he actually caught sight of what he believed to be the tip of the continent. He named it the Cape of Storms, reflecting the turbulent conditions at the junction of the Atlantic and Indian Oceans. On Dias's return home King João II renamed it the Cape of Good Hope, so as not to discourage the next waves of explorers.[19]*

Dias' voyage would change the course of history. First, he had confirmed that the classical geographer Ptolemy was wrong and that there *was* an end to Africa; thus a sea-route from

* King João II was given by his great rival, Queen Isabella of Castile (who later unified Spain) perhaps the greatest accolade of history. She referred to him only as 'The Man'[20] – a moniker far better than even Bruce Springsteen's …

Europe to the riches of the Indian Ocean circumventing the Islamic world was eminently plausible. But second, and just as importantly, he had discovered the band of westerly winds in the South Atlantic that can reliably carry mariners round the tip of the continent.[21] Rather than hugging the African coastline and battling against northerly currents after crossing the equator, the solution is instead to steer a wide looping course into the mid Atlantic. The same *volta do mar* trick developed for returning from the Canary Islands in the North Atlantic also works in the South Atlantic – the wind bands in the Northern and Southern hemispheres are mirror images of each other. This gave European navigators the first inkling of the grand-scale patterns of circulation in the planet's oceans and atmosphere, which they soon came to understand more deeply and began to exploit.

A NEW WORLD

While the Portuguese had been finding a route around the southern tip of Africa, a Genoese navigator was trying to raise support for a voyage sailing in the opposite direction: he believed he could reach the orient by sailing west. He finally found patronage from Queen Isabella of Castile, who in 1469 had married King Ferdinand II of Aragon to unify their realms and form Spain. He was known to his sponsors as Cristóbal Colón. In English, we call him Christopher Columbus.

Contrary to a commonly held view today, no educated person in medieval Europe believed the Earth to be flat. In the third century BC Eratosthenes, a Greek geographer, astronomer and mathematician working at the Library of Alexandria, understood that the world is a sphere and calculated its circumference to be 250,000 stadia, or around 44,000 kilometres – remarkably close to its real value. Indeed, the techniques of celestial navigation used by sailors to plot their latitude by the stars is predicated on the very principle that the Earth is round. Columbus was also not the first person to propose that India could be reached by sailing west from Europe: the Roman geographer Strabo had suggested the same in the first century AD. And there was evidence that something did lie over the watery horizon. Reports from the Atlantic islands described flotsam drifting in from the west: unfamiliar trees, canoes and the bodies of people neither European nor African in appearance.[22]

In order to secure financial support for his expedition, Columbus had to convince potential sponsors that the proposed voyage was achievable. But how do you estimate the distance travelling west from the edge of Europe to China or India before you've accomplished such a journey? The solution was to start with the calculated circumference of the world and then subtract the overland distance from Europe to the Orient – the approximate width of Eurasia was known from travellers along the Silk Road. The problem was that these calculations yielded a westward oceanic distance of around 19,000 kilometres, or about four months of sailing with reliable winds. Such a journey was utterly impossible at the time. Ships simply couldn't carry enough food and clean water to keep their crews alive for this long on the open seas without making landfall for fresh supplies.

Not to be deterred, Columbus pulled the sort of sleight of hand used by any intransigent believer absorbed in the strength of their own convictions. He fudged the numbers. Columbus took the lowest calculation of the Earth's circumference available at the time, along with the greatest estimate of the breadth of Eurasia, and arrived at a significantly shortened sea distance to the west. He used the measurements of Paolo dal Pozzo Toscanelli, a Florentine mathematician and map-maker, who not only grossly underestimated the circumference of the globe, making the world a third smaller than it really is, but also believed that Japan lay 2,400 kilometres east of China, thus offering a chance to break up the long sea voyage. Columbus argued that he would be making landfall in the islands off Japan after travelling only 3,900 kilometres from the Canary Islands. This would amount to just a month's sailing time.

Indeed, Columbus claimed that the Orient lay not far over the horizon from the position of the Azores.[23] He never considered the possibility of an unknown continent lying in the way: by his calculations there simply wasn't room for one in the western seas.

The Portuguese refused to sponsor the venture, however. King João II's advisors regarded Columbus' numbers as a dangerous underestimate and the proposition foolhardy. And in any case, Bartolomeu Dias had just successfully rounded the Cape of Good Hope and shown Portugal the open gate to the Indian Ocean via their African route. The Genoese, Venetians and English also declined to get involved; but at last Columbus' repeated lobbying of the Spanish court bore fruit. Queen Isabella was advised that although the proposition might be high-risk, it also offered the possibility of enormous gains. And this is where a degree of blind historical luck played to Columbus' fortunes.

The Treaty of Alcáçovas in 1479, which ended the War of the Castilian Succession, had handed the Canary Islands to Castile while the Portuguese kept possession of Madeira, the Azores and the Cape Verde Islands. This agreement worked clearly in favour of the Portuguese in the Atlantic, with Castilian ships forbidden to sail to these archipelagoes; indeed, the Portuguese were given exclusive rights over any lands that had been or would be discovered south of the Canaries. If Castile wanted to pursue its own territorial and trading interests the captains would have to head west. And it just so happens that the Canaries offered the ideal starting point for ships sailing in that direction across the Atlantic.

Had Columbus' proposal been accepted by King João II he would have embarked on his audacious westward voyage from the Azores. Lying as they do about 850 kilometres further west of Madeira and the Canaries, we now know that these islands are situated about a third of the way from the edge of Europe to the American coast. But the Azores also lie further north than the other Atlantic islands, and at this latitude the prevailing winds blow towards the east – unfavourable for any Atlantic crossing. The Canaries, however, lie within the zone of the northeasterly trade winds that blow all the way to the Caribbean. By sheer historical fluke, Isabella's support – and the Treaty of Alcáçovas – meant that Columbus attempted his crossing from an archipelago that happens to be upwind of the Americas. If his expedition had set sail from the Azores it would likely have perished deep in the ocean.[24]

On 3 August 1492 Columbus' three ships slipped their moorings at the port of Palos de la Frontera and sailed south-west to the Canary Islands. Here Columbus restocked his provisions, made some repairs, and then turned the prow of his ships in the direction of the sunset. Carried by the easterly trade winds over the featureless expanse of the Atlantic Ocean, they made landfall five weeks later in the Bahamas.* Columbus then continued further south-west to explore the coastlines of Cuba and Hispaniola. Here he heard of a people inhabiting the Lesser Antilles island arc whom the Spanish named *cariba* or *caniba*, from which we get our words Caribbean and cannibal.†

After four months of exploring these islands, Columbus was ready to return home to receive his expected riches and glory. But how to travel back from a place that has never before been reached by sea? Columbus first tried to simply sail back the way he had come, but he soon realised that his ships would have a hard time beating a course upwind into the same easterlies that had borne them on the outward journey, and that they risked running out of provisions before reaching land. He decided to turn north instead, and in the middle latitudes, he picked up the same band of westerly winds that blow past the Azores and which carried him back to Europe. Therefore

---

* In this way, it took Columbus just over a month to sail across the ocean that had taken more than 100 million years to open by plate tectonics.

† He also learned from the natives of the West Indies about their hammocks, which would change how European sailors slept on board ship for hundreds of years.

Columbus' expedition would have been impossible without the knowledge gained by Portuguese sailors that the prevailing winds blow in opposite directions in neighbouring bands of latitude, won through their methodical efforts to push down the African coast for decades before Columbus had even been born.[25] Crossing the Atlantic in midwinter exposed the weary sailors to ferocious storms, but after a month's sailing Columbus' ships arrived safely in the Azores,[26] and from there they returned to Spain.

Columbus made a total of four voyages west, mapping out the chains of tropical islands in the Caribbean, but it wasn't until the third expedition that he actually set foot on the American mainland, in what is today Venezuela. Yet to the end of his life, Columbus still maintained that he had reached the Orient.[27]

By the early 1500s, scores of tropical islands had been mapped by European sailors, as well as the long South American coastline continuing past the equator with its big rivers, suggesting that they drained an expansive area inland. Other explorers also reported large land masses far to the north. Alarmed by Spain's supposed new route to Asia along the latitude of the Canary Islands, England's king Henry VII sent the Venetian navigator Giovanni Caboto (or John Cabot) on an expedition to find an alternative route through the North Atlantic, which reached Newfoundland.

It became clear that Columbus had not reached the Orient – but what exactly had been discovered? It began to dawn on Europeans that perhaps the lands to the west were all one continuous coastline, that they had stumbled across not a series of new islands, but an entire continent – a whole New World.

THE GLOBAL WIND MACHINE

The Portuguese had spent the best part of a century inching their way down the coast of Africa before they finally found its southern tip and the gateway into the Indian Ocean. Now, within a generation of the discovery of the Americas in 1492, European sailors ventured across all the world's oceans and completed the first circumnavigation of the Earth. This was a revolution that heralded the birth of today's global economy.

All this was only possible because mariners had come to understand the patterns of reliable winds and currents around the globe, which now determined the trade routes that brought great riches to Europe. But what causes these alternating bands of prevailing winds around the world, which in turn drive the great wheeling currents in the oceans?

The warmest part of the Earth is the equator, which receives the most direct sunlight over the year. The air near the equatorial surface heats up and rises, but as it ascends it cools, the moisture condenses into clouds and then falls as rain. At high altitude, the cooling air mass diverges and splits to the north and south, like a T-junction high in the atmosphere. Each of these arms travels about 3,000 kilometres before descending to the ground again, now very dry, at around 30° latitude – roughly one-third of the way between the equator and the pole – in both the Northern and Southern hemispheres. These two bands around the Earth are called the subtropical highs because the air crushing down here creates slightly higher pressure. The warm air rising from the equator, on the other hand, leaves behind a low pressure region.

From the subtropical highs at 30° latitude the air then travels back towards the equator as surface winds to complete this great vertical circuit. This zone of reliable winds, which was so important for the European crossing to the Americas, is a manifestation of the same atmospheric circulation pattern that produces the world's great bands of tropical rainforests and mid-latitude deserts which we discussed in the last chapter. These two immense atmospheric rolling patterns, convection currents just like those around your radiator at home, are known as Hadley cells, and they operate like paired cogs, separated by the equator and rotating in opposite directions. The movement of the Hadley cells, driven by equatorial warming, is a great heat engine – no different in principle from a steam engine or the internal combustion engine in your car, albeit one with a power rating of about 200 million million watts[28] – ten times greater than the entire power use of global human civilisation today.

But there's another important aspect of our planet that influences the winds on Earth. Our planet, and its atmosphere, is rotating. Because the Earth is a solid sphere, this means that the surface at the equator is moving faster than that at higher latitudes. And so as the air returns from the subtropical highs towards the equator the ground beneath turns eastwards faster and faster. There's a small amount of friction between ground and atmosphere that starts to drag the air along with the surface, but the air is unable to pick up speed sideways fast enough as it moves, and so the winds blowing towards the equator get left behind by the rotating surface. The upshot is that they are effectively deflected in a smoothly curving path towards the west. This is known as the Coriolis effect, and it influences anything moving over the surface of a rotating sphere, for instance the trajectory of ballistic missiles. Or to put it another way, if you imagine yourself rocking about on a ship in tropical waters, the prevailing winds appear to blow from the east, but a more accurate description would be that you and the Earth's surface are spinning quickly through the atmosphere, and the easterlies are like the wind in your hair when you're driving fast in your car with the top down.

Any winds blowing in the Northern Hemisphere are deflected by the Coriolis effect to their right, and those in the Southern to their left. Thus between the latitude of 30° north and the equator the prevailing winds follow a curved path towards the south-west, and so by the nomenclature for winds are called northeasterlies. And the same is true in the Southern Hemisphere: the air returning north along the surface to the equator is again deflected towards the west to produce prevailing southeasterly winds. These easterlies are known as the trade winds, and as reliable winds blowing through the tropics they have been absolutely crucial for mariners.*

The band where the returning northeasterly and southeasterly trade winds meet each other around the equator is called by modern atmospheric scientists the Intertropical Convergence Zone (ITCZ). But to sailors it's known as the doldrums. This is the region of low-pressure air, characterised by light winds or periods of dead calm, that was first encountered by Portuguese sailors crossing the equator on their journey down the African coast in the late fifteenth century. The region can prove disastrous for ships as they wait for the winds to pick up again or an ocean current to carry them out. They can find themselves becalmed and stuck in the doldrums for weeks, and in this equatorial region of hot and muggy climate it may mean not only a delay in delivering your cargo back to port but it can also spell death as your onboard fresh water supplies run out. Samuel Taylor Coleridge evoked the desperation of sailors becalmed in the Pacific doldrums in 'The Rime of the Ancient Mariner':

> Day after day, day after day,
> We stuck, nor breath nor motion;
> As idle as a painted ship
> Upon a painted ocean.
>
> Water, water, every where,
> And all the boards did shrink;
> Water, water, every where,
> Nor any drop to drink.

The location of the ITCZ is determined by rising air warmed by the sun, and so it shifts north and south of the geometric

---

* Surprisingly, though, the name for these winds does not derive from the meaning of 'trade' in the sense of commerce. The term in fact originates from a different sixteenth-century usage: a wind 'blowing trade' means it is in a constant direction. Thus the trade winds are constant, proving to be very useful for exploration and trade in our understanding of the word.

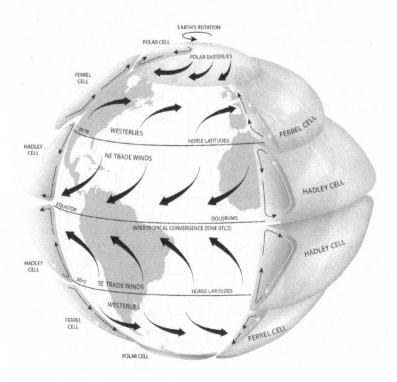

The grand circulation currents in the Earth's atmosphere that create the alternating bands of prevailing winds.

line of the equator with the seasons. And because land warms up more quickly than the ocean in summer, the ITCZ band is pulled further away from the equator by the continents. It therefore follows a decidedly sinuous, snakelike path around the waistline of the world. This makes the exact location and width of the ITCZ hard to predict, and increases the risk of sailors getting caught out in the doldrums.

Beyond the 30° latitude of the descending arms of the Hadley cells, at around 60° north or south the surface air, although cooler than at the equator, is still warm enough to rise into the atmosphere and drive another convection loop. And just as with the Hadley cells, the surface winds blowing back towards the equator at the bottom of this loop are smoothly deflected to their right by the Coriolis effect, producing the band of winds called polar easterlies.

The third and final pair of grand circulation currents in the Earth's atmosphere are the two Ferrel cells, operating in the middle latitudes between 30° and 60°. But unlike the other two, the Ferrel system is passive: it's not directly driven by its own rising warm air, but by the rolling of the Hadley and Polar cells it nestles between. It's almost like a freewheeling gear being forced round by two powered cogs turning on either side of it. Where the descending arms of the Ferrel and Hadley cells merge, at around 30° north and south, they form two subtropical ridges of high pressure known as the 'horse latitudes'. These regions are also characterised by light, variable winds or calm conditions; and so, like the doldrums, sailors learned to be wary of them.

Because the Ferrel cell is driven by the Hadley and Polar cells on either side it turns in the opposite direction. And this fact has been hugely important in the Age of Sail. The surface winds of the Ferrel cells blow not towards the equator but the poles, and so the Coriolis effect instead deflects them in the opposite direction. This is the zone of the westerlies. Two different latitude bands of winds blow to the west – the trade winds of the Hadley cell and the polar easterlies – but if you want to sail east you can only do that within the realms of the two Ferrel cells and the westerly surface winds they produce. This is the route back to Europe from Central and North America, first exploited by Columbus once he realised he needed to sail north into this zone to return home.

The zone of westerly winds has proved just as vital in the Southern Hemisphere. As we noted previously, due to a quirk of the current distribution of the continents through plate tectonics, the Northern Hemisphere is packed with land masses and their mountain ranges which disrupt the flow of winds. The Southern Hemisphere, on the other hand, is dominated by open ocean, free of windbreaks. In particular, below about 40° only the bottom tip of South America and the two islands of New Zealand impede the uninterrupted rush of the westerly winds all the way around the world. The southern westerlies consequently tend to be much stronger than their northern counterparts, and sailors came to call this zone the Roaring Forties. And if they dared to push even further south, risking fierce wind and waves, frigid climate and threat from icebergs, navigators could take advantage of the even stronger Furious Fifties or Shrieking Sixties.

This pattern of alternating bands of winds between the equator and the poles also drives the currents in the world's oceans which too have been enormously important for knitting together our world into huge trade networks. The neighbouring zones of the easterly trade winds and the westerlies blow the surface water in opposite directions. This, coupled with the fact that the continents block the water from simply circling the world, and that water moving north or south across the globe also feels the Coriolis effect, creates great wheeling surface currents known as ocean gyres. There are five major gyres, in the North and South Atlantic, North and South Pacific, and in the Indian Ocean. These ocean gyres turn clockwise in the northern hemisphere, and anticlockwise in the southern, and like the direction of the wind bands they mirror each other over the equator.

The Canary Current, coursing along the North African coast, was well known to Phoenician and later Iberian sailors, as we saw earlier. This is the eastern arm of the gyre circulating in the North Atlantic; the Gulf Stream, bearing warm waters from the Caribbean up to Northern Europe, forms the western arm. The Gulf Stream was discovered in 1513 when Spanish explorers sailing south along the coast of Florida realised they were being pushed backwards despite sailing with a strong wind. (Because water is so much denser than air, even a gentle ocean current can have a much greater effect on a sailing ship than the wind.) The commercial implications were immediately realised: heavily-laden galleons needed only slip into this wide, fast-flowing river within the ocean to be readily carried north and then round with the westerly winds back home.[29] The Brazil Current, running along the east coast of South America, is the mirror-image counterpart of the Gulf Stream and carries ships south into the zone of the westerly winds which they then pick up for rounding Africa into the Indian Ocean.*

So overall, in each hemisphere the atmosphere enveloping the planet is divided into three great circulation cells, like giant tubes wrapped around the world, each rolling in place and shifting north and south slightly with the seasons. These produce the major wind zones of the planet – easterly trade winds, westerlies and polar easterlies – which in turn drive the circulating ocean currents. Pretty much the entire wind pattern on Earth can therefore be explained by three simple facts: the equator is hotter than the poles, warm air rises, and the world spins.

This sums up the general pattern of banded winds around the globe. But there is one region of the world with a unique wind system that drove a thriving maritime trading network long before Europeans encountered it.

INTO THE MONSOON SEAS

When you hear the word 'monsoon' your mind may well fill with imagery of a verdant and muggy Indian landscape being lashed by torrential rain of fat, heavy drops. The word derives from the Arabic *mausim*, meaning 'season',[30] and the monsoons are of course crucial for the wet and dry seasons that shape

---

* The action of the fluid dynamics in these vast rotating gyres of ocean is to pull surface material into the centre of the gyre. The Sargasso Sea lies in the middle of the North Atlantic Gyre – and is the only region of open ocean to be classified as a sea – forming a 1,000- by 3,000-kilometre patch of distinctively clear, blue water that is thick with seaweed. The same corralling process has in recent times also concentrated large amounts of plastic flotsam, dubbed the North Atlantic Garbage Patch; a similar concentration of pollution is found in the Pacific Trash Vortex.

agriculture across South East Asia. But scientifically speaking, the monsoons are the result of the distinctive atmospheric conditions around South Asia and the pronounced rhythmical reversals in the direction of the prevailing winds. Here was a system of winds completely alien to anything Portuguese sailors had encountered before in the Mediterranean or the Atlantic.

Following in the footsteps (or at least the ship's wake) of Bartolomeu Dias, another Portuguese explorer, Vasco da Gama, set sail from Lisbon in the summer of 1497 to complete the sea route to India. He took the now customary route along the north-west African coast, re-watered in the Cape Verde islands, and then travelled round the bulge of Africa. But rather than hugging the familiar African coastline into the doldrums in the Gulf of Guinea, he turned his ships south-west into the gaping expanse of the Atlantic, enlarging Dias' *volta do mar* into a huge looping course that took him thousands of kilometres away from land.[31] Far out at sea he encountered the Brazil Current that carried him steadily south until he picked up the prevailing westerly winds, discovered by Dias a decade earlier, which bore him readily back east to the tip of Africa.

Da Gama and his crew had just spent over three months at sea, travelling around 10,000 kilometres through the Atlantic, making it by far the longest voyage through open ocean undertaken by that time. By comparison, Columbus had spent only 38 days sailing west before his nervous crew became mutinous and demanded to turn back – only to fortuitously sight land two days later.

Da Gama now worked his way round the cape, pushing against the current that sweeps around the south-eastern African coastline. On 16 December 1497 they passed the final stone pillar erected by Dias. By the following March, he reached Mozambique, entering the realm of the Arabic sea traders. In the port of Malindi, in modern-day Kenya, he first encountered Indian merchantmen and here da Gama was able to secure the services of a Gujarati pilot with knowledge in navigating the

Indian Ocean.[32] Heading off in late April they were blessed with a steady wind to the north-east – da Gama was yet to appreciate the nature of the monsoon winds and the fortuitous timing of his journey – and the fleet headed on a diagonal course cutting right across the Indian Ocean, making for Calicut on the Malabar Coast. On 29 April they noticed the North Star on the horizon: they had re-entered the Northern Hemisphere. Vasco da Gama's ships arrived in Calicut on 20 May 1498, after having crossed over 4,000 kilometres of open ocean in only twenty-five days. He had finally achieved the decades-long dream of Portugal's explorers and traced a sea route from Europe to India and the riches of the Spice Islands.

The Portuguese spent some time exploring the Indian coast before embarking on the voyage home in early October. But now it showed that de Gama's grasp of the rhythmical mechanics of the monsoon wind patterns was woefully inadequate: no navigator with local knowledge would have attempted to cross south-west to the African coast at this time of year. Da Gama's ships found themselves battling against a headwind and were forced to beat back and forth, making very slow progress. Worse still, they were frequently becalmed, while their drinking water turned foul and scurvy reared its ugly head amongst the crew.*

They eventually reached the East African coast at Mogadishu. Their dismally mistimed return passage had taken 132 days. If only they'd waited two months before attempting the journey they could have sailed before the winter monsoon winds to

---

* It was during these first long-distance voyages by the Portuguese that scurvy began to routinely afflict sailors. Scurvy was not unknown at the time: it occurred during times of famine or among armies on nutritionally unbalanced diets. But it was sailors voyaging the seas for months on end that the disease now affected with great regularity, indeed inevitability. Today we know that scurvy is caused by vitamin deficiency. Vitamin C, or ascorbic acid, is a vital ingredient in the way the body makes collagen for the connective tissue. Within a month or so of a diet containing insufficient vitamin C, the symptoms progressively worsen, ranging from bleeding gums and

make the crossing in just a few weeks. By the time the Portuguese finally made it home they had been away for almost exactly two years and travelled around 40,000 kilometres.[34] Their feat of courage and endurance had come at the cost of the lives of two-thirds of the crew, many having succumbed to scurvy. The rhythms of the monsoon winds must be heeded.

But their ships had returned with holds full of cinnamon, cloves, ginger, nutmeg, pepper, and rubies, whereas Columbus' first expedition had found little of any worth. So although it was Columbus' eight-month expedition of 1492 that is most remembered today, in many ways da Gama's 1497 voyage was far more impressive. He had discovered what Columbus set out, but failed, to find: the sea route to the riches of the East.

THE MONSOON METRONOME

The monsoon winds are driven by exactly the same process you'll be familiar with from the changing breezes on a trip to the seaside. During the day, the land warms up more quickly, and to a higher maximum temperature, than the surface of the sea alongside it. This causes the air over the land to rise, and the cooler air over the sea is sucked into the low pressure region left behind, driving a convection current with a steady wind that blows from the sea to the land – an onshore breeze. Conversely, the ground cools much more quickly after sunset, and so the warmer, rising sea air

---

aching bones to poor wound healing, loss of teeth and eventually convulsions and death. Curiously, humans are one of the few animal species (guinea pigs are another) that suffer from scurvy. It turns out that at some point during our evolutionary divergence from other primate species we picked up a mutation in a single letter of our genetic code that knocked out the key enzyme for making ascorbic acid in our own liver cells. Scurvy was the major killer of sailors on long voyages until the end of the eighteenth century, when it was identified that citrus fruit could prevent the disease.[33]

pulls in behind it air from the land to drive an offshore breeze. If you sit on the beach over sunset you can often feel the distinct reversal in wind direction. The only difference is that monsoons occur on a much grander scale, and seasonally rather than daily. In the summer the land mass of continents heats up more quickly than the surrounding sea surface, driving a monsoon wind that pulls in moist air from over the ocean. Through the winter, the ocean retains more of its warmth so that the convection cell reverses, the monsoon winds flip direction, and dry air from higher in the atmosphere descends onto the continent.

Seasonal monsoon winds are created by the temperature differences between several continental land masses and their surrounding ocean. West Africa as well as North and South America also experience weak monsoons but the monsoon winds over India and South East Asia are by far the strongest on Earth, and this comes down to geography. The Tibetan Plateau is the world's largest and highest, measuring roughly 2,500 by 1,000 kilometres, and rising on average more than 5 kilometres above sea level. When the ground of the Tibetan plateau warms in the summer sunshine it also heats up the air in the upper atmosphere. This gives a big boost to the rising air currents at the beginning and end of the summer monsoon season. Even more important in driving the strong monsoon winds is the Himalayan mountain range along the southern edge of the plateau. This acts like a high wall, a barrier that blocks cold, dry air from the north being sucked down over India and mixing with the warm, moist air from the ocean, which would subdue the atmospheric circulation. The Himalayas essentially insulate India and provide the conditions for a powerful monsoon effect.[35] So the strong monsoon winds of south Asia are another consequence of plate tectonics – the result of India crashing into Eurasia about 25 million years ago.

India sits in its enveloping ocean like the central spike in a huge 'M', and as it warms up with the beginning of summer, the rising air currents suck in moist air from the ocean that encircles it, which then itself ascends, cools and condenses into clouds that release huge amounts of monsoon rain. The Intertropical Convergence Zone, as we saw earlier, snakes around the waistband of the Earth where the trade winds blowing from the north and south meet each other. During the summer, the heating of India and the effects of the Tibetan Plateau and the Himalayas are so pronounced that the ITCZ is pulled over 3,000 kilometres north of the equator, and then swings far to the south again in winter. Thus the ITCZ band sweeps through the region, the trade winds from the Southern Hemisphere push right over the equator in summer, and in winter the northerly winds extend into the Indian Ocean and islands of the East Indies.

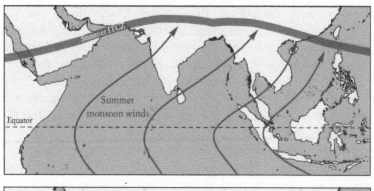

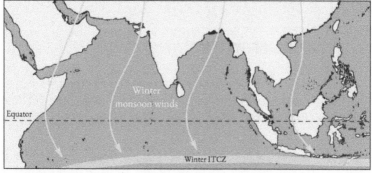

The seasonal pattern of reversing winds in the Monsoon Seas.

In effect, the geography of India disrupts the 'normal' wind patterns that we encounter around the rest of the globe. From one season to the next, the winds over the whole of South East Asia periodically flip direction, like the great breathing in and out of huge planetary lungs. Through the eleventh to fifteenth centuries, long before the arrival of Portuguese mariners, ships making use of these winds to sail across the Indian Ocean and among the myriad islands of the East Indies created a dynamic and diverse trading network, with bustling ports along the routes.[36]

The reversal of the monsoon winds is as regular and predictable as a metronome, and simply by timing your voyages right you can sail to where you need to go on favourable winds, load up on goods and re-provision your ship, and then simply wait for the winds to shift direction and carry you all the way home again. Navigating in the Indian Ocean or around the East Indies is therefore different from sailing the Atlantic or the Pacific. There the trick is to move north or south between neighbouring atmospheric circulation cells with either the tropical easterly trade winds or the mid-latitude westerlies – you pick the wind direction you need by a change in space. But the trick for navigating the monsoon seas is to wait for the seasonal reversal and sail back pretty much the way you came – you pick the right wind direction by a change in time. And this is something that Vasco da Gama completely failed to appreciate when he entered the Indian Ocean in 1498.

EMPIRE OF WATER

From the year after da Gama's return, the Portuguese began sending annual expeditions to India along his new route.* These mariners also learned their lesson from da Gama's punishing

---

* The first fleet that followed da Gama's route to India took such a wide *volta do mar* loop through the south Atlantic on the way out that they discovered Brazil.[37]

return voyage and quickly acquired knowledge of the rhythms of the monsoon winds that dictate the sailing schedules through the Indian Ocean and islands of South East Asia. Now in possession of this key navigational understanding, and with their large, cannon-wielding ships and experience of building strong fortifications born of centuries of incessant warfare in Europe, the Portuguese rapidly asserted their dominance in the region and continued further east in their quest for the source of the spices. In 1510 they conquered Goa, turning it into their main base of operations around the Indian Ocean,* and the following year they took Malacca so as to control the maritime traffic through these straits. Once they had ascertained where the Spice Islands were located, they sent an expedition to occupy the Moluccas in 1512. The Portuguese also gained permission to establish trading centres in Macau, on the south Chinese coast, in 1557, and in Nagasaki, Japan, in 1570.

By 1520, the income from Portugal's spice trade across the Indian Ocean provided nearly 40 per cent of the Crown's total revenue. Portugal had created a new kind of empire, made powerful and wealthy not through possession of large areas of territory but by the strategic control of sprawling oceanic trade networks on the other side of the world – an empire of water.[39]

Where the Spanish and Portuguese had led the way, the Dutch, British and French followed. The rivalry between these marine trading powers triggered colonial wars around the world as they attempted to eject each other from strategic ports and forts, and control chokepoints to dominate the critical sea passages. Through exploration and maritime trade Europe's centre of gravity shifted decisively from east to west. Europe was no longer the world's western extremity, the distant terminus of the network of the Silk Road threading its way across Asia.

* When Sri Lankans first encountered the Portuguese and their alien European food and drink they reported that 'They eat a sort of white stone and drink blood.' It was the first time they had seen bread and wine.[38]

And the Mediterranean – the inland sea that had witnessed millennia of city states, kingdoms, and empires competing with each other for dominance – became almost parochial, fading from its previous centrality into relative insignificance.

The New World, and the new maritime routes to India and the Orient, offered Europeans access to a seemingly inexhaustible trove of territory and resources, wealth and power. As European navigators decoded the secrets of the planet's wind patterns and ocean currents they reached across the great expanses of the world's oceans, linked formerly unconnected regions of the planet, and began the process of globalisation.[40] The Age of Exploration was therefore not just a process of filling in the world map with strange new lands, but also of discovering invisible geographies. European sailors learned how to use the alternating bands of planetary winds and wheeling ocean currents like a great interlinked system of conveyor belts, to carry them where they wanted to go.

The early exploration ships were slender-hulled and rigged for the greatest manoeuvrability around unknown coastlines, and in particular for beating into the wind. But these small caravels with triangular 'lateen' sails required large numbers of expert crew and had little stowage space for cargo alongside the necessary provisions. The ideal design for transoceanic trade is a broad ship rigged with large square sails, which is much simpler to handle and so minimises the crew size whilst maximising the hold space for supplies and profit-making cargo. These square-rigged ships, exemplified by the Spanish galleons, catch a great deal of motive force but can only ever ride with the wind: beating against the breeze is virtually impossible.[41] This meant that in contrast to the early years of exploration, the trade routes that came to establish the European imperial presences overseas were strongly dictated by the direction of the prevailing winds, and this had profound implications for patterns of colonisation and the subsequent history of our world. The three most important of these were the Manila Galleon Route, the Brouwer Route and the Atlantic Trade Triangle.

TOWARDS GLOBILISATION

While the Portuguese were establishing their trade empire in South East Asia, the Spanish were exploring their possessions in the Americas and began looking for their own westward route to the riches of the Spice Islands.

By 1513 a Spanish explorer had trekked across the Panama isthmus and was the first European to set eyes on the ocean on the far side.[42] As we saw in Chapter 2, Ferdinand Magellan – a Portuguese navigator but sailing for the Spanish – passed round the bottom tip of South America in 1520 through the strait that now bears his name and called this new ocean the 'Mare Pacificum' – the Peaceful Sea.[43] His fleet rode north along the coast with the Humboldt Current of the South Pacific Gyre before picking up the trade winds to sail westward to the Philippines, which he claimed for Spain. Magellan was killed on the island of Mactan, but his fleet continued on their voyage and in 1521 reached the Moluccas, the famed Spice Islands themselves and at the time the sole source in the world of nutmeg and cloves.*

The problem for the Spanish voyage to the Spice Islands was that the mariners had found a route west across the Pacific, but did not know the winds needed to return east to the Americas.

---

* The two great marine powers of the time had signed a treaty in 1494 to split the world between a Portuguese East and a Spanish West. The division, known as the Tordesillas line, ran north–south through the Atlantic 370 leagues (just over 2,000 kilometres) west of the Cape Verde Islands. It was no more than a line on a map passing through featureless, open ocean – a pure carto-graphical abstraction. When Portuguese sailors on their way to India discovered the coast of South America they realised it lay on their side of the demarcation and so claimed it: which is the reason why Brazil speaks Portuguese and the rest of Latin America Spanish. The problem that arose in the 1520s was what happened around the other side of the planet. If the Tordesillas line is extended in a circle all the way across the poles and through the Pacific – 180° opposite the Atlantic partition – do the Moluccas lie in the Spanish or Portuguese realm? In the event the dispute was resolved when Spain, in urgent need of quick cash to finance its ongoing war with France, sold its Moluccas claim to Portugal.[44]

The only ship of Magellan's expedition to make it home did so by continuing west across the Indian Ocean, completing the first circumnavigation of the globe. As her captain wrote: 'We made a course around the entire rotundity of the world – that by going by the Occident we have returned by the Orient.'[45]

It took another forty years before Spanish sailors had gained knowledge of the winds that allowed a return journey back east across the Pacific to the Americas. Realising that the pattern of winds in the Pacific replicates that of the Atlantic, navigators sailed north from the Philippines as far as the coast of Japan before picking up the band of westerlies (in the Ferrel circulation cell of the atmosphere) that carried them in the right direction.[46] The discovery enabled the Spanish to bridge the vast Pacific Ocean with regular round-trip shipping, the Manila Galleon Route. This ran between the colonies of New Spain in Acapulco, in present-day Mexico, and Manila in the Philippines, and for 250 years – from 1565 to 1815, ending with the Mexican War of Independence – this Pacific crossing was the longest-running trade route in history.[47] The westerly winds across the Pacific delivered the galleons to the coast of California, where they needed way stations to be resupplied after this long ocean crossing and before setting off on the last leg of their journey south down the coast to Mexico. This explains the strong Spanish colonial presence in the region, with the names of the major cities of San Francisco, Los Angeles and San Diego still recalling the Spanish influence today.

The main cargo carried west across the Pacific on this route was silver. In the 1540s the Spanish discovered rich silver veins in Mexico, as well as the 'silver mountain' of Potosí high in the Andes.[48]* Most of this silver was taken up the South American

---

* Potosí, also known as Cerro Rico (Spanish for 'rich mountain'), is the core of an eroded volcano[49] formed about 13 million years ago. The volcanic activity drove an underground hydrothermal system that leached out silver, as well as tin and zinc, from deeper rocks and then redeposited it in extremely rich, thick veins[50] riddled throughout the heart of the mountain. It is the largest silver mine in history and for more than 100 years accounted for over half of global production.[51]

coast on the Humboldt Current to the Panama Isthmus, carried across this narrow land bridge by packs of mules and then loaded onto ships bound for Spain.[52] Sailing across the Atlantic in treasure flotillas the Spanish galleons were the prey of French, Dutch and English corsairs with memorable names like 'Peg Leg' Le Clerc and Francis Drake.

About a fifth of the mined American silver was sent across the Pacific aboard Manila Galleons and in the Philippines it was traded for Chinese luxuries: silk, porcelain, incense, musk and spices.[53] Ultimately, whether it was carried on the Manila Galleon route to the Philippines for trade with the Chinese, or back to Spain and then filtered through the European empires towards the east, about a third of all South American silver flowed to China,[54] which placed a higher value on this precious metal even than on gold. Some of the silver was traded with India, where in the early seventeenth century the Mogul ruler Shah Jahan built a resplendent mausoleum for his wife – the Taj Mahal. This enduring symbol of love also epitomises the early global economy that was taking off with the Age of Sail: South American silver exploited by the Spanish, and handled through European merchants, ultimately financed a monumental building project in India.[55]

For a time, Spain grew immensely rich and powerful with this stream of silver coming out of the Americas. But like the Atlantic Trade Triangle, to which we'll come later, this immense European wealth came at a great human cost for the workers mining the depths of the silver mountain for months at a time, suffering the heat and dust at a lung-straining altitude of 4,000 m. Potosí has been memorably described as the 'mountain that eats men'.[56]

The seventeenth century saw another crucial new route opened to the East Indies. The passage discovered by the Portuguese in the last years of the fifteenth century went round the tip of Africa, followed up the continent's eastern coastline and crossed to India, before heading round to the Strait of Malacca. The route dipped only slightly into the band of westerly winds to carry ships past the southern tip of Africa. These are the Southern-Hemisphere

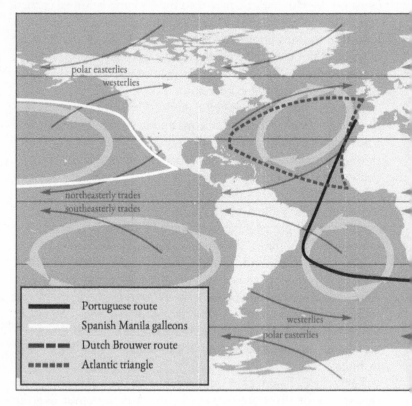

Connecting the world: major oceanic trade routes exploiting different wind bands and ocean currents.

mirror image of the mid-latitude westerly winds that the Spanish learnt to ride from the Philippines to Mexico on the Manila Galleon Route. But as we saw earlier, in the Southern Hemisphere these westerlies are unobstructed by major land masses and are consequently far stronger. It wasn't for over another century, however, that sailors realised how they could fully exploit the Roaring Forties.

In 1611, Captain Henrik Brouwer of the Dutch East India Company passed the Cape of Good Hope and instead of heading

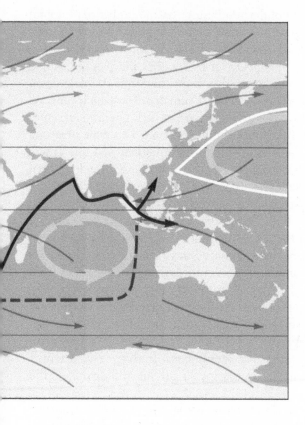

north-east towards India turned south, deeper into the westerlies. These carried him fully 7,000 kilometres east before he exited this fast-moving ocean freeway and turned north again to Java. The Brouwer Route, making use of the Roaring Forties, took less than half the time of the traditional passage – not least because it obviated the need to wait for the monsoon winds in the Indian Ocean. As well as offering a far quicker voyage to the Spice Islands, this more southerly and cooler route away from the tropics kept the crew healthier and supplies fresher.

The development of the new passage had a number of profound historical consequences. It was sailors taking the Brouwer Route who first set eyes on the west coast of Australia. And detouring south around the Indian Ocean meant that the passage shifted the gateway into the East Indies from the Strait of Malacca to the Sunda Strait between Java and Sumatra. The Dutch founded Batavia – present-day Jakarta – in 1619 as their operational centre in the region and to command this key strait. This zone of strong winds was also the reason behind the founding of Cape Town: the Dutch needed a resupply port for ships before the long final leg of their journey. The Roaring Forties wind belt is therefore the reason why Afrikaans is spoken today in South Africa. *

It was spices that drove the early years of the Age of Exploration and the global oceanic trade carried by European ships, but by 1700 new commodities had come to dominate demand. Crops originally grown in Africa and India had been transplanted to the New World and large amounts of coffee were now being produced in Brazil, sugar in the Caribbean, and cotton in North America.[58] And the demand for labour needed to mass-produce these commodities for the European markets led to another transcontinental trading system, which is arguably the most significant of all for the shape of the world today.

In simple terms, the Atlantic Trade Triangle linked Europe, Africa and the Americas to serve Europe's insatiable hunger for cheap cotton, sugar, coffee and tobacco. Ships sailed from Europe with goods manufactured in these developed nations,

---

* One major problem throughout most of the Age of Sail was that ships' captains had a hard time determining their precise position in the open seas. Astronomy can easily tell you your latitude – you only need to measure the angle between the horizon and select stars – but before the invention of accurate clocks it was nigh-impossible to work out your correct longitude. Ships dashing east along the Roaring Forties had to know the right moment when to turn north-east to continue up to Indonesia. And if you waited too long you would plough into Australia – the continent's coral-studded western coastline is littered with the wrecks of ships that missed their turning.[57]

such as textiles and weapons, down to the West African coast to trade them with local chiefs for slaves they had captured. They then transported these slaves across the Atlantic to sell them in the colonies, to plantation owners in Brazil, the Caribbean and North America.* The capital raised by selling this human cargo was used by the captains to purchase the commodities grown on the plantations, the produce of the slaves' labours. The cargo holds of the slave ships were scrubbed with vinegar and lye and they then took these raw materials back to Europe for manufacturing, and so completed the loop.[61] There were variations on the exact routes sailed and the wares transferred at each leg in overlapping circuits, as well as short hops shuttling goods along certain stretches of coastline,[62] but this was the core of the Atlantic Trade Triangle that operated between the European homelands and their colonial territories from the late sixteenth to the early nineteenth centuries.

Before shipment across the Atlantic, African slaves were held in coastal forts, known as factories, often established at the mouth of rivers as this offered the easiest way for transporting captives from further inland. The great majority of slaves were taken from West Central Africa – the region between the equator and about 15° south – and along the Gold Coast, the Bight of Benin and the Bight of Biafra in the Gulf of Guinea. This too is largely due to the mechanics of atmospheric circulation patterns and ocean currents. From these locations it is easier to cross to South America with the southeasterly trade winds and then south down the coast with the Brazil Current for the Brazilian coffee plantations; or follow the northeasterly trades and north equatorial current to the sugar plantations of the Caribbean islands, the cotton

* The Portuguese had begun importing African slaves to the sugar plantations on Madeira and the Cape Verde Islands in the late 1400s, and from the 1530s transported them across the Atlantic to their colonies in Brazil.[39] Soon enough, other European seafaring nations got involved in the human trafficking of the so-called Middle Passage.[60]

plantations of Alabama and Carolina, and tobacco plantations of Virginia. The Atlantic slave trade was banned in 1807, but continued by smugglers until the abolition of slavery with the conclusion of the American Civil War in 1865. By this time over 10 million Africans had been forcibly seized and transported to the Americas,[63] many dying in the abysmal conditions on the way or in the first year or two on the plantations. About 40 per cent were taken to Brazil, 40 per cent to the Caribbean, 5 per cent to what became the United States, and 15 per cent to Spanish America.[64]

The shipping merchants sold their cargo for a profit at each stage of the triangle, and so like an economic perpetual motion machine the system generated huge financial gains for its masters with each turn of the crank. While the European nations began to use waterwheels and then steam engines to power their mills and factories, the enslaved human workforce overseas providing the raw materials was an equally important component of the machinery driving the economics of industrialisation. Before the forces of abolitionism gathered strength, the taste of sweetened tea or a slug of rum, the feeling of a clean shirt on the back, and the invigorating inhalation of pipe smoke made Europeans close their minds to the human suffering that was ultimately providing for their lifestyle.[65]*

The huge areas of new land that made up Europe's overseas colonies and the raw materials and profits they provided helped create the conditions for the Industrial Revolution, but just as crucial for powering this transformation was the availability of seemingly limitless amounts of energy from the subterranean world, to which we'll now turn.

---

* And we're no more responsible consumers today, excitedly buying the latest electronic touch-screen device or cheap T-shirt whilst knowing in the back of our minds the appalling conditions many factory workers in the developing world are forced to endure.

Chapter 9

Energy

For the vast majority of the 10,000-year history of settled humanity we have been an agrarian society. Sedentary peoples have fed themselves on crops grown in the nearby fields and raised animals for meat, milk and traction power. Husbandry has also provided the fibres we make into clothes to protect ourselves from the elements: cotton, linen, silk, leather and wool.

In essence, agriculture gathers solar energy from a certain area of land, and transforms it into nutrition for our bodies and raw materials for our community. Over time we have increased the agricultural output either by expanding the area under cultivation – by clearing forest to make way for farmland and developing new tools and techniques, such as the heavy plough, to cultivate previously marginal land – or by selective breeding of higher-yielding crops and animals, and crop-rotation schemes. We have become increasingly adept at this through history, and consequently our population has boomed.

Felling forests has also provided the firewood we need for cooking our food and heating our homes. And timber provided the heat energy needed to convert the raw materials we gathered from the natural environment into products like pottery, bricks, metals and glass. To create the greater temperatures required in our kilns, furnaces, forges and foundries, we have carbonised wood to make charcoal. In this way, by relying on charcoal made from forests, even steel and glass production was tied to the growth of trees.

As our population grew and the demand for timber to be used as fuel and construction materials increased, we began to run out of nearby natural forests and learned how to coppice. Coppicing is the system of managed forestry where trees such as ash, birch and oak are felled and allowed to resprout from the trunk to redevelop into another mature tree. Coppicing can be carried out in repeated cycles to provide a continual supply of wood from the land.*

But as the population in Europe continued to grow, even coppicing could not satisfy our insatiable appetite for firewood and construction timber. From the mid seventeenth century this shortage became more and more acute and the price of wood rose inexorably. Europe was hitting 'peak wood': all suitable land was already being used to grow food and the production of fuel could not be increased any further. Then, however, a new source of energy began to be explored that not only kept our home fires burning, but provided levels of energy that far outstripped muscle power.

SUNSHINE AND MUSCLE POWER

For most of human history, the power required for building and maintaining civilisation had been provided by muscles, whether those of human labourers or draft animals. Muscles, properly employed and coordinated, can achieve phenomenal feats: the pyramids of Giza, the Great Wall of China, the cathedrals of medieval Europe – all were constructed with muscle power and simple mechanical contraptions like rollers, ramps and winches. But muscles need to be fuelled by food, which in turn requires farmland and pasture. And so as our

---

* Many of the tree species of Northern Europe – including alder, ash, beech, oak, sycamore and willow – are able to resprout from a snapped stem, and it is this natural capability that makes them suitable for coppicing. But this in itself may have developed as an evolutionary response to damage wrought by foraging elephants and other megafauna – the sort of huge animals that roamed across even northern latitudes in the warmer interglacial periods, as we saw on page 33.[1]

population swelled and agricultural land became increasingly scarce, muscles got expensive.[2]

There already were alternatives to muscle power that harnessed natural, renewable energy sources. Much was accomplished with the turning force provided first by the waterwheel and then the windmill. The waterwheel was invented around 2,500 years ago, and by the first century AD it was used by the Chinese for driving the bellows in blast furnaces for smelting iron.[3] The most extensive waterwheel facility constructed by the Romans was in Barbegal in southern France, built soon after AD 100.[4] Here a system of sixteen waterwheels made up the greatest known concentration of mechanical power in the ancient world,[5] with a total output equivalent to 30 kilowatts of power.* Windmills first appeared in Persia in the ninth century AD, and were constantly refined as they spread across medieval Europe. The Low Countries in particular adopted windmills with great enthusiasm for draining polders and reclaiming land from the sea, as we saw in Chapter 4. Waterwheels and windmills came to provide the motive power for everything from grinding grain into flour, pressing olives for oil, sawing timber, crushing metal ores and limestone, and driving rollers to squeeze iron bars into shape.

This mechanical revolution, which gathered pace from the eleventh to the thirteenth centuries, saw medieval Europe become the first society not to base its productivity on the straining toil of human or animal muscles alone. But even so, energy availability remained a limitation on productivity, with waterwheels and windmills subject to the vagaries of river levels and wind breezes. Even though waterwheels or windmills alleviated the physical exertion of running production processes, we continued to live in a world run on muscle power and sunshine.

Over history we have learned to divert the energy of the sun through the ecosystem and channel it instead into our bodies

---

* While this was hugely impressive for its time, it is still utterly dwarfed by the prodigious amounts of energy we have learned to marshal today: this entire waterwheel complex is outstripped by the power output of a single family car engine.

and our society. It was sunshine that ripened our crops and nurtured our forests. Indeed, for most of our history the productivity of civilisation had relied on, and was limited by, photosynthesis and how quickly plants could generate food and fuel on the land we had at our disposal.

This system has been given various labels, such as the Organic Energy Economy, the Somatic Energy Regime, or the Biological Old Regime,[6] but they all allude to the same truth: before the eighteenth century, the entire history of civilisation had been supported by solar energy harvested by crops and forests, and muscle power supplied by human labourers and draught animals which in turn had to be fuelled by food gathered from plants. But if society's productivity is governed by the growth rate of crops and coppices – by how quickly you can harvest the sunshine – it is fundamentally limited by the suitable land that is available. Moreover, the food you eat and the firewood you need for manufacturing compete with each other over the same land. There is a hard ceiling on what agricultural empires can achieve.

The only way to escape these limitations is to find sources of energy that don't require you to harvest sunlight directly. And this was accomplished in eighteenth-century Europe, by tapping into the huge stockpiles of stored energy lying under our feet. Rather than trying to extract more energy from the land surface, we burrowed underground to extract caches of ancient forest growth, in the form of coal. Coal is essentially a combustible sedimentary rock and a single coal seam represents the condensed essence of many seasons of forest growth – it is fossilised sunshine. Just one tonne of coal can provide as much heat energy as a year's firewood taken from a whole acre of coppiced woodland. It was coal that built the modern world.

THE POWER REVOLUTION

We used coal long before the Industrial Revolution. When Marco Polo travelled along the Silk Road to China in the late thirteenth century, he described how the Chinese had the odd practice of burning pieces of black stone for fuel.[7] And even in Britain, by the end of the second century AD, the Romans were mining many of the main coalfields in England and Wales for use in metalworking or their underfloor heating systems.[8]

It was textile manufacturing that set in motion the process we call the Industrial Revolution. A series of inventions transformed this cottage industry in the second half of the eighteenth century, with machines now able to spin cotton and wool fibres into thread, and then weave these threads into fabrics. The availability of cheap cotton from Britain's colonies in America and India – we've explored these international trade networks in the last chapter – supplied this growing demand as the mills rapidly increased their capacity, and at first waterwheels provided the motive power. But the force that really drove the progression of the Industrial Revolution was the virtuous circle that existed between coal, iron production and the steam engine.

The Industrial Revolution started gathering momentum with the introduction of coke to fuel blast furnaces. Coal dug from the ground is not pure carbon fuel, but contains impurities like volatile organic compounds, sulphur and moisture. Coking is the process by which coal is first heated without allowing it to ignite and burn – in much the same way as charcoal is produced from wood – to drive out these impurities to create a hotter-burning fuel, and in particular to remove the sulphur which can taint the iron and make it brittle. Coke-fired blast furnaces made the production of iron much cheaper, providing the material for construction projects and increasingly sophisticated machine tools.

The exploitation of huge underground coal reservoirs, and the coke produced from it, released early industrialising Britain from the limitations of coppiced forests and provided an enormous supply of energy for creating the products needed by society. But it was the steam engine that marked a truly monumental advance, providing force and movement without the need for animal muscles. Fundamentally, the steam engine is a converter, able to turn thermal energy into kinetic energy: it transforms heat into motion. The first steam engines were employed at coal mines to pump out groundwater so that ever deeper seams could be dug. Given that they were sited at collieries, it didn't matter that the earliest, primitive designs were enormously hungry for fuel. But a string of innovations and improvements made steam engines increasingly energy-efficient and powerful.

The steam engine became a general-purpose power plant. It served as the 'prime mover' in factories, where a single engine could drive a whole workshop floor of machine tools via a system of overhead belts and chains. More compact and fuel-efficient, high-pressure steam engines were developed for transport, their considerable weight spread out across the surface by the laying of railway tracks; or they were mounted in a ship, supported by the buoyancy of the hull. Steam soon came to haul freight and passengers around the world. By 1900, steam engines provided about two-thirds of the power needed in Britain, carried 90 per cent of all land transport along railways and bore 80 per cent of cargo across the seas.[9]

This was the essence of the three-way process that drove the accelerating pace of industrialisation. Steam allowed us to mine ever greater amounts of coal, coal-fired smelters and foundries produced more and more iron, and both coal and iron were used to construct and run more steam engines to mine coal, produce iron, and build yet more machinery at ever increasing rates. In this way, coal, iron and the steam engine formed a virtuous triangle

The reason why this industrial transition is so important in our history is that it released us from the previous energy limitations on human civilisation. Coal provided prodigious amounts of thermal energy without the need for coppicing, and the steam engine removed the reliance on animal and human muscles. Without huge reserves of buried fuel it is unlikely that civilisation would ever have progressed beyond an essentially agrarian state. So how has the Earth provided this ready-to-go energy resource waiting for us?

FOSSILISED SUNSHINE

You will no doubt know that coal was created by the burial of ancient trees. And as we have seen repeatedly throughout this book, once again there was something quirky about the geological period which saw the most productive and widespread era of coal formation. These prevailing conditions had profound ramifications for life on Earth.

Although plants first colonised the land about 470 million years ago, evolving from branching green algae growing in lakes,[10] it took a long time for plant cover to build up sufficiently to produce the earliest, and still very minimal, coal deposits. Within the almost 400 million years when substantial forested areas covered our planet, by far the most massive and widespread coal deposits were created in the Carboniferous, a 60-million-year period ending about 300 million years ago. Indeed, it is coal formation that gave this geological age its name. There have been other, later periods of coal formation in our planet's history, but the Carboniferous dominates by the sheer quantity of coal deposited and its widespread extent. Around 90 per cent of the coal we've used since the Industrial Revolution dates to this short period of geological history.

Normally, when living organisms die, whether that's an oak or an owl, they are decomposed to release the carbon in the organic molecules of their bodies back into carbon dioxide in the air, which is then captured again by photosynthesising plants. For such vast amounts of carbon to be turned into coal during the Carboniferous, something has to block that decomposition process, and it seems that in that period for some reason the Earth's carbon recycling scheme broke down. Trees died but didn't rot. Fallen vegetation accumulated on the ground, becoming peat, which was then buried deeper and deeper underground to be cooked into coal in the internal heat of the planet.

The key prerequisite for peat to accumulate is simply that the growth of vegetation has to be faster than the rate at which the dead material can be removed by decomposition or, on a longer timescale, the deposits can physically erode away. And it was lush, vigorously growing forests in a low-lying, subsiding swamp environment, where dead trees became buried without oxygen before they could decay completely, that seems to have tipped the balance.

In the Carboniferous Period, our world looked very different. This far back in time, the layout of the continents, constantly scuttling around the face of the Earth under the influence of plate tectonics, was in a completely different configuration. Throughout this period, the major landmasses were crunching together to be welded into a single whole, assembling the super-continent Pangea.

Great, low-lying basins in what is now eastern North America and Western and Central Europe sat across the equator, forming tropical swamps where dense forests flourished. The trees that filled these swamp forests still reproduced with spores – as described in Chapter 3 – and would have looked unsettlingly alien to us. They were ancient relatives of the horsetails, club-mosses, quillworts and ferns that in today's forests humbly occupy the shady understorey. Much of the coal that eventually formed was produced by lycopsids,[11] trees related to today's clubmosses. Their metre-thick trunks were very straight with few side-branches, curiously green-coloured, and textured with a regular dimple pattern where old leaves had fallen off – fossils of these trees look almost like tyre tracks. Growing over 30 metres high, they supported a compact crown of long, blade-like leaves.

These lush wetland ecosystems also teemed with grotesque animal life. The Carboniferous undergrowth trembled with giant cockroaches remarkably similar in appearance to those of today, spiders the size of horseshoe crabs (although not yet spinning webs), and 5-foot millipedes. Newt-like amphibians as large as horses[12] lumbered through these swamps with a wide-limbed, sprawling gait. And giant predatory dragonflies, with wingspans up to 75 centimetres,[13] soared through the hot and humid air. But if you could travel back in time to take a stroll through these lush forests you'd be struck by the notable absence of certain sounds which would feel eerie once you noticed it: the complete lack of birdsong. In these ancient skies only insects had taken to the air – birds would not appear for another 200 million years. Many other creatures you might expect in this sort of environment were also yet to evolve – no mosquitoes whined among the tepid pools, but there were also no ants, beetles, flies or bumblebees.[14]

While the Carboniferous provided the ideal environmental conditions for lush tree growth, later eras were also warm and muggy, so these alone cannot account wholly for the huge coal deposits from this period. It is less their vigorous growth than the fact that the dead trees didn't rot away and accumulated in thick layers of peat that requires explanation. The languid swamps around the Carboniferous equator would certainly have helped, the oxygen-poor soil in these fetid marshes slowing the activity of decomposition microbes. But swamps have existed throughout planetary history: they're not a uniquely Carboniferous feature.

So what might have been special about the world around 325 million years ago? Why did fallen tree trunks seem so reluctant to rot away? Why did the carbon recycling fail so spectacularly during the Carboniferous, leading to the creation of much of the coal that was to fire the Industrial Revolution?

One explanation, which has become popular over recent years, is that in the Carboniferous fungus, which plays a central role in the decomposition process, simply wasn't biochemically equipped to break down the fallen trees.

To grow ever taller, the early trees needed to develop greater internal strength to support themselves. All plants contain cellulose, a molecule made up of long chains of sugar units that strengthens their cell walls – a linen jacket, a cotton shirt and the paper page you're reading at the moment (unless you're swiping through on an e-book device, in which case think of the cardboard box it came in) are all composed of cellulose.

But what really gave these towering trunks their strength was the biological invention of a second molecule: lignin. It explains why the small moss-like plants of the early Devonian became the towering trees of the Carboniferous. And importantly, lignin is much harder to break down than cellulose.[15]

If you walk through a forest today, your nose filling with the heady aromas of the humus-rich soil and the foliage, you'll notice that the dead wood of a log to the side of the path has become pale-coloured, and soft and spongy in texture. This is caused by white rot fungi, which decompose the dark-coloured lignin in wood. (Particularly tasty varieties include the oyster mushroom and shiitake.) But during the Carboniferous Period, so this theory goes, trees had newly formulated lignin to reinforce their wood, but fungi had not yet had time to develop the necessary enzyme toolkit to break it down. Much of the solid bulk of trees had become indigestible and for millions of years when they fell they simply piled up on the ground.

But while this is a satisfying hypothesis, it unfortunately doesn't stand up to more recent evidence. For one, the most common kind of coal-forming trees in the Carboniferous swamps didn't actually contain much lignin. And although North America and Europe didn't form much coal in the geological period immediately following the Carboniferous – the Permian – some areas in China did, and this was after the supposed emergence of lignin-decomposing fungi.[16] So if it wasn't an evolutionary time lag between forests reinforcing themselves with lignin and fungi developing the capability to digest it, what was it about the Carboniferous that made it so very prolific at turning trees into coal?

It appears that the reason for the vast Carboniferous coal deposits is not primarily biological but geological.

While the tropical region around the equator remained warm, the late Carboniferous Period was actually a pretty cold time in the Earth's history, with large ice sheets forming in the south of Gondwana – so despite its popular conception, the Carboniferous world was not all steamy jungles. These glacial conditions were caused by the configuration of the continents at the time. The congregating land masses stretched from the South Pole, across the equator and pretty much all the way to the North Pole. This blocked the circulation of warm tropical and cold polar ocean waters around the world – the conveyor belt we encountered in Chapter 8 – obstructing the transfer of heat from the equator to the poles. The fact that Gondwana sat across the South Pole also supported the accumulation of thick glacial ice in the region: as we saw, ice caps cannot grow as extensively over open ocean.

The vigorously growing Carboniferous forests were also partly responsible for triggering these glacial conditions.[17] The trees had been sucking carbon dioxide out of the air for their photosynthesis, and when they died and much of their organic material became locked up as peat rather than rotting away, they didn't release that carbon back into the air. As a result, the amount of carbon dioxide in the atmosphere dropped considerably, and the low levels of this greenhouse gas would also have contributed to global cooling. And as the decomposition of dead organisms consumes oxygen from the air to make that carbon dioxide, an increase in peat production also led to an increase in atmospheric oxygen levels, perhaps to as much as 35 per cent[18] (today the concentration of this life-sustaining gas in our atmosphere is 20 per cent). These high oxygen levels are believed to have contributed to the evolution of the giant insects, such as the large-winged dragonflies, we encountered earlier.

From the mid Carboniferous, therefore, Earth was becoming an icehouse. Fluctuations in the global temperature and thus the amount of water locked up in the ice caps (governed by wobbles in the Earth's orbit, as we saw in Chapter 2) caused cycles of rising and falling sea levels, just as during the ice ages of the past 2.5 million years. As the Carboniferous sea rose and fell it repeatedly advanced and then retreated over vast stretches of the low-lying swamps. In the process, vegetative matter got buried regularly under layers of marine sediment, to one day become coal seams. Indeed, if you look at the exposed layers of rocks in a coal measure you'll see a vertical sequence of coal seam, marine sediments like mudstones, lagoon sediments such as shale, and then sandstones from a river delta on which a new soil layer has formed, followed by another coal seam. These stacked layers within the coal measure can be read as a geological manuscript telling the story of the repeated flooding of the swamp basins.

In places like South Wales or the Midlands of England, where coal is found next to ironstone, both fuel and ore for smelting can be sourced in the same place – it's like the Earth is offering us a 2-for-1 discount. Sometimes we even get a 3-for-1 deal. Lying just beneath the coal measures and so often found exposed on the surface in the surrounding landscape, are limestones formed during the early Carboniferous when global ocean levels were high and flooded low-lying land in shallow, warm seas. As we saw in Chapter 6, limestone is used as a flux in iron smelting to help the metal melt and remove impurities. Furthermore, the 'seatearth' layers just beneath each coal seam, which often preserve the fossilised roots of the swamp trees, are frequently rich in hydrous aluminium silicates. Such minerals make this clay layer exceedingly refractory and able to withstand high temperatures of 1,500 °C or more, so Nature also provides us with the perfect fireclay building material for lining the furnaces or crucibles for pouring molten metal.[19] Thus the changing conditions through the Carboniferous sometimes provided in the same area in successive strata the raw materials for the Industrial Revolution.

It was this periodic flooding and burial of the low-lying swamps that preserved the peat and compressed it under successive layers of sediment to form coal. And the swings between glacial low and interglacial high sea levels, reflecting the icehouse conditions at the time, were a direct consequence of plate tectonics and the configuration of the world's continents. But there was a second unusual feature of our planet during the Carboniferous which contributed to coal formation: the landmasses weren't just arranged in a clump between the North and South poles but were still actively crashing together.

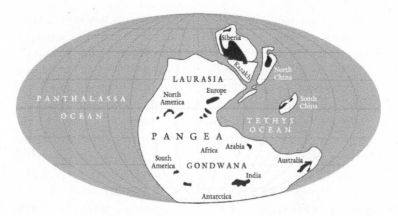

Major coal-forming basins during the construction of the supercontinent Pangea.

The Carboniferous Period saw the ongoing construction of the supercontinent Pangea, as the great northern continent, Laurasia (containing North America and parts of northern and western Eurasia), collided with Gondwana (South America, Africa, India, Antarctica and Australia) along the equator. This slow crunching event is known as the Variscan Orogeny* and it created a thick belt of mountain ranges, including the Appalachians along what is now the Eastern Seaboard of the US and Canada, the Lesser Altas Mountains in Morocco – which would have continued from the Appalachians before this huge mountain range was separated by the opening of the Atlantic Ocean – and many other ranges across Europe such as the Pyrenees between modern-day France and Spain.† Then, in the Late Carboniferous, Siberia slid from the north-east into this continental pile-up, welding onto Eastern Europe and producing the Ural Mountains at the join.

---

* Orogeny is a geological term for the building of a mountain range from subduction or collision of tectonic plates, although disappointingly the adjectival form is 'orogenic' not 'orogenous'.

† The orogeny also created the intrusions of granite in Cornwall, which as we saw came to supply tin for bronze production and kaolinite clay for making porcelain.

As we have seen, the collision of continents not only creases up chains of high mountains, but creates low-lying, subsiding basins alongside as the crust flexes downwards. A good example is the Ganges Basin sitting along the foot of the Himalayas, formed by the Indian-Eurasian plate collision, which holds the Indus and Ganges rivers as they flow down from the mountains to the sea.

Such down-warping foreland basins were also created by the tectonic clashes of the Carboniferous and provided the setting for those huge areas of low-lying swamps that were prone to periodic flooding and so smothered and preserved the peat. But for coal deposits to build and not be eroded away through exposure as the cycle of deposition continues, you need a basin that is also continually subsiding. And this is what's so important about the ongoing formation of Pangea during the Carboniferous: continental collisions kept the basins warping down at roughly the same rate that the coal built up, allowing enormously thick successions of coal seams to accumulate.

It is this chance coincidence of several factors all acting at the same time and place that made the Carboniferous such a unique period in Earth's history for creating the massive deposits of coal that we came to rely upon. The Pangea supercontinent was still being actively constructed with clashing frontiers that just so happened to be situated around the tropics, creating foreland basins for low-lying wetlands in a warm, humid climate perfect for tree growth. These swamps were repeatedly flooded with sudden sea-level rises during a rare period of oscillating glacial and interglacial ages, burying and preserving the peat; and they were continually subsiding so the strata weren't simply eroded away again. The process of plate tectonics was the ultimate force behind all this. There were to be later periods of coal formation around the world, but none would be as productive as during the Carboniferous assembly of Pangea.[20]

This confluence of planetary factors ultimately fuelled the Industrial Revolution. Without the great Carboniferous coal measures humanity might have stalled in its technological development three centuries ago. We might still be using waterwheels and windmills, and tilling our fields with horse-drawn ploughs.

THE POLITICS OF COAL

There were many reasons why the Industrial Revolution began in Britain. The scarcity and rising price of wood (and thus charcoal) encouraged the substitution of coal as a fuel wherever possible. The economy of labour in Britain favoured the replacement of expensive craftspeople with machines which, although they required high initial capital investment, were more productive and required fewer workers to operate. And Britain's empire provided cheap cotton from America and then India that prompted innovations allowing textiles to be more rapidly produced from fibres. So although in Britain the introduction of machines replaced human labour, it was slaves toiling in fields overseas who produced the raw materials like cotton that drove the process.

But Britain also benefited from a geological bonanza – mountains of easily accessible, good-quality Carboniferous coal lying underground waiting to be dug up – to literally fuel its industrialisation. By the 1840s, Britain's coalfields were supplying so much energy that to match it using charcoal instead would have required burning 15 million acres of woodlands[21] – an area equivalent to a third of the entire country – *every year*.

The Industrial Revolution spread from its birthplace as tools, techniques and technologies for intensive coal mining and the mass production of iron and steel were adopted in mainland Europe. Here the same formation of Carboniferous coal that had fuelled Britain's industry extends underground through northern France and Belgium to the Ruhr region of Germany. This was to become the industrial heartland of Europe: a coal crescent as central to modern history as the Fertile Crescent was to the ancient world.[22] In North America, the transition to coal occurred much later: the far less densely populated colonies along the East Coast initially had access to huge areas of forest for charcoal,[23] so American industry did not begin to replace charcoal with coal on a grand scale until the mid nineteenth century.[24] Nonetheless, by the 1890s

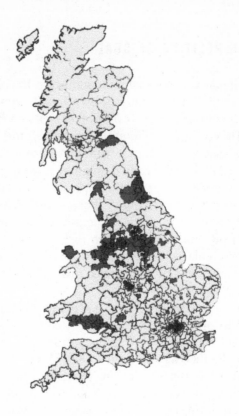

The 2017 UK election map (*above*) with Labour constituencies shown in dark, and the Carboniferous coalfields (*right*).

the US had surpassed Britain as the world leader in the production of iron and steel.[25] In particular, Pittsburgh was well positioned within easy reach of iron ore, limestone for the flux, and the abundant coal measures of the Appalachian Mountains – a geological concurrence that made the fortunes of some of the richest tycoons of the modern capitalist era, such as Andrew Carnegie.

Today the collieries that fuelled the Industrial Revolution in Britain have virtually all closed, as remaining coal seams became

ever more difficult to access, cheaper coal became available overseas, and less polluting or renewable energy sources were sought.* A few opencast mines remain, but the last of the deep mines in the UK, Kellingley in Yorkshire, closed in 2015.[27] Yet astonishingly, the distribution of the 320-million-year-old coal-fields in the UK still leaves its imprint on Britain's political map today.

The Labour Party was founded in 1900 from the trade union movement, with particularly close ties to British coal miners.

---

* On 21 April 2017, Britain went a full day without using coal to generate electricity for the first time since the 1880s.[26]

And although it has changed a great deal over the last hundred years – from a party in the shadow of the Liberals to a landslide victory just after the Second World War to New Labour under Tony Blair – the deep link between coal and politics has endured for generations. Take the results of the most recent General Election, in 2017, for example, shown on page 270. The election was much closer than the map would seem to imply, with densely populated, multicultural cities such as London tending to lean to Labour and the sparsely populated, larger, rural constituencies overwhelmingly voting Conservative. The result was a hung parliament with Labour winning 262 seats in parliament and the Conservatives 318 – not enough for a majority government.

But let's look more closely at the distribution of Labour votes across the country. The figure on p. 271 shows the location of Britain's coalfields, and what is remarkable is the tight correlation between the political and geological maps. The broad coalfields of Cumberland, Northumberland and Durham (the Great Northern), Lancashire, Yorkshire, Staffordshire, North and South Wales, all match perfectly with the large areas of Labour constituencies in the election. This correlation was even stronger in the 2015 election, when Labour's crushing defeat confined it to its heartlands, and the pattern is evident throughout the previous decades. Support for the main left-wing political party in the UK almost perfectly matches the regions of Carboniferous deposits.* It seems that the old geology deep underground is still reflected in people's lives today.

While coal remains a crucial part of the world's energy mix, primarily for generating electricity and manufacturing steel and concrete, the politics of coal have now been largely superseded by those of another fossil fuel. Today, oil is one of the most valuable commodities in the world and the dominant energy

---

* The correlation between Labour and coalfields is less clear in Scotland with the rise of another major left-wing party, the Scottish Nationalist Party.

source of humanity, making up a third of the total consumed by our global civilisation.[28] Geopolitical tensions over its production and transport have dominated international relations for decades, and as we saw in Chapter 4, it is the major reason for the West's interest – and interventions – in the Gulf and the naval chokepoints around the world through which the oil supertankers must pass.

BLACK DEATH

Like coal, we have used petroleum – literally, 'rock oil' – for millennia. Asphalt (bitumen) that had seeped up onto the surface was used as a cement in the construction of the walls of Babylon 4,000 years ago and as a road-building material around 625 BC.[29] By AD 350 the Chinese were drilling oil wells and burning the fuel to evaporate brine to produce salt,[30] and in the tenth century Persian alchemists were distilling petroleum to make kerosene for lamps. But it was not until the second half of the nineteenth century that we began to use oil on an industrial scale.

Crude oil is a hugely complex mixture of different-sized carbon compounds, which can be separated by distillation into different fractions. Early uses of these fractions included lubricants for steam engines and other machinery, and kerosene for lighting cities. But it was with the development of the modern internal combustion engine in Germany in 1876 that humanity's oil consumption really took off.[31] The gasoline refined from crude oil had previously been considered too volatile and dangerous to be of much use, but it proved a perfect fuel for powering the pistons of these new machines. Today we also use aviation kerosene for soaring over the clouds in our planes. The long hydrocarbon compounds in these liquid fuels pack in far more energy than coal and so represent fabulously dense and portable stores of power for transport.[32] And not only does oil fuel our automobiles, it is also crucial to laying down the smooth roads on which we drive – viscous asphalt is composed of the longest hydrocarbon chain molecules in crude oil.[33]

Oil is so attractive because it ranges high on the Energy Return on Investment (EROI) index – that is, you only need to put a small amount of energy into extracting and refining it, but you'll get a huge amount of energy out of it.[34] It is also far more portable than coal: the liquid crude can simply be squirted down pipes over huge distances. It is this winning combination of high energy density, easy transportability and relative abundance that has turned oil into the most important source of energy in the world today. And it's not just critical as a fuel. Around 16 per cent of its annual production is not burned but used as feedstock for a diverse range of organic chemistry, producing everything from solvents, adhesives and plastics to pharmaceuticals. Modern intensive agriculture would also be impossible without oil. It is used to synthesise pesticides and herbicides that control the artificial environment of farms for high yields, it fuels the tractors and harvesters that tend the fields, and artificial fertilisers are also made by using fossil energy. Oil feeds your car but you're also drinking it with every meal.

While coal was produced by the Earth from the compacting and baking of ancient swamp forests, oil and natural gas are formed from the remains of microscopic marine plankton. Life has been flourishing in the seas for far longer than plants have colonised the land masses, but most of the oil powering our twenty-first century civilisation was actually formed about 200 million years after the Carboniferous forests flourished. This oil was created in the now-vanished Tethys Ocean in two huge pulses about 155 and 100 million years ago:[35] during the late Jurassic and mid Cretaceous periods.

The sunlit surface waters of the world's oceans today are teeming with microscopic life made up by hordes of tiny critters collectively known as plankton. The primary producers forming the foundations of the oceanic ecosystems are phytoplankton like diatoms, coccoliths and dinoflagellates. These single-celled photosynthesisers grow by absorbing the energy of sunlight to capture carbon dioxide and fix it into sugars and all the other organic molecules they need; and just like land plants they release oxygen as a by-product. While the Amazonian rainforest is often referred to as the lungs of the planet, in fact it is the drifting multitudes of phytoplankton in the seas that produce most of the oxygen we breathe. And when conditions are just right for their growth, staggeringly dense populations of these cells amass in the water – the milky turquoise blooms of coccoliths are even visible from space.

The planktonic realm is also filled with zooplankton – microscopic grazers and predators such as forams and radiolarians. These microorganisms are able to extend little tentacles through pores in their elaborately shaped hard shells to ensnare and devour less fortunate plankton. Both phytoplankton and zooplankton are in turn eaten by fish, which are eaten by larger fish, or filtered out of the water in huge gulps by whales, and so they form the foundation of the entire oceanic food web. When the plankton cells evade their predators and die of natural causes, they are consumed by decomposition bacteria that recycle the carbon and other elemental nutrients back around the system. This planktonic ecosystem of primary producers, predators, scavengers and decomposers is as complex as the Serengeti with its grasses, gazelles, cheetahs and vultures, but all played out in microscopic miniature in the sparkling surface waters of the world's oceans.

When plankton die they drift down through the water column to darker and darker depths, joined by slowly sinking mineral grains blown in by wind or washed in by rivers from the continents. This steadily settling precipitation of decaying organic matter and inorganic detritus towards the sea floor is known as marine snow. Even the deepest depths of today's oceans are well oxygenated by the global circulation of seawater, and so most of the organic remains are digested by bacteria and the carbon recycled.

This is what happens across the great majority of the oceanic expanse today. But to accumulate organic debris on the seafloor, which eventually turns into oil, you need high planktonic productivity in the surface waters combined with limited oxygen at the ocean bottom, to prevent bacteria recycling the carbon so that it instead accumulates as a black, organic-rich mud on the seafloor (analogous to the conditions needed to build coal seams, as we saw earlier). This carbon-loaded mud then becomes buried under further deposits so that it is squashed and hardened into black shale rock. This is the starter material for crude oil and natural gas around the world. As the shale becomes buried deeper and deeper it is warmed by the interior heat of the planet, until it passes into what is known as the 'oil window' – a temperature range of about 50–100 °C. Simmering slowly, the complex organic compounds of the dead marine life are broken down into the long-chain hydrocarbon molecules of oil. If the shales are exposed to higher temperatures, up to about 250 °C, the deep chemistry breaks apart even these long chains into small carbon-containing molecules, mostly methane, but also some ethane, propane and butane – that is, natural gas. The oil window generally occurs at a depth of between 2 and 6 kilometres, and it can take the shale over 10 million years to become buried this deep by the continuing sedimentation above.

The enormous pressures at this depth squeeze the liquid oil out of its source rock and back up through the overlying strata. If it doesn't encounter anything to block its vertical migration and hold it underground the oil simply seeps back out from the sea floor. Sandstone works very well as a reservoir rock, the pore spaces between the individual grains acting to soak up the oil like a geological sponge, and with a layer of, say, finely grained mudrock or impermeable limestone on top to act as a seal, the oil and gas becomes trapped, ready for us to drill down and suck it up.[36]

As we've seen, this process no longer happens in our oceans today. So what were the peculiar conditions in the ancient Tethys Ocean 100 million years ago that caused so much plankton debris to accumulate and become oil?

By the Cretaceous Period, the great supercontinent Pangea had fragmented and the continents were dispersing again. No longer was there a single massive landmass draped across the equator. Instead, the huge waterway of the Tethys Ocean stretched all the way around the midriff of the world, separating the northern and southern continents. This meant that the ocean circulation patterns were very different back then, with a current able to flow unhindered in a circuit around the whole world. This equatorial current bathed in the tropical sunshine and became very warm.

Indeed, the mid Cretaceous world was a broiling hothouse, with equatorial sea surface temperatures as high as 25–30 °C, and still a tepid 10–15 °C at the poles. No ice caps existed there, and Canada and even Antarctica supported dense forests. Without ice caps locking up large amounts of water, the sea levels were also much higher than they are today. In addition, lots of active rifting took place in the Earth's crust at the time, opening up the North and South Atlantic as the continents pulled apart. As new oceanic crust is created in these sea-floor spreading centres it is still warm and buoyant and the crust bulges up in great long ridges of submarine mountain ranges. These huge mid-ocean ridges displaced a great deal of water and the sea levels rose even higher. Indeed, the combination of hot climate and active sea-floor spreading meant that sea levels were higher during the late Cretaceous than at any other period over the past billion years of our planet's history – they were perhaps as much as 300 metres higher than today.[37]

Consequently, the ocean inundated huge areas of the continents: Europe was mostly submerged; the Western Interior Seaway flooded right up through the middle of North America from the Gulf of Mexico to the Arctic (as we saw in Chapter 4 when we looked at the voting patterns in the south-eastern US); and the Trans-Saharan Seaway swept down Africa from the Tethys through what is now Libya, Chad, Niger and Nigeria. Vigorous volcanism associated with the widespread rifting also released a lot of nutrients into the seas to fertilise plankton blooms. The late Cretaceous was therefore a world not just of deep ocean but also of shallow marginal seas, whose warm waters provided ideal growth conditions for plankton.

But conditions were also very different on the floor of these Cretaceous seas. In a hothouse world without polar ice creating

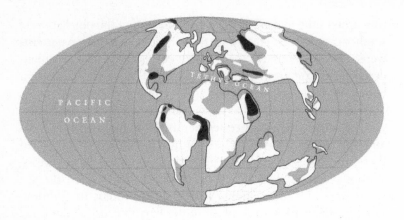

Oil-forming regions in the anoxic seas of Cretaceous Earth.

cold, dense water, the thermohaline circulation that we explored in Chapter 3 was shut down: there was no global conveyor belt circulating water through the ocean depths. And crucially, warm water also holds much less dissolved oxygen, and any that did make it into the deep waters was readily used up by the decomposition bacteria.

The upshot of all this was that the Cretaceous sea floor became an oxygen-starved dead zone, where the bacteria weren't able to properly break down the organic matter. At the same time, the frantic productivity of the plankton in the warm, sunlit surface waters produced a veritable blizzard of marine snow settling down to the sea floor. Without being decomposed, the organic material accumulated and became buried as more sediment settled down.* As with the Carboniferous coal forests in subsiding swamp basins, the carbon recycling system in the Cretaceous sea floor had become broken, allowing organic matter to accumulate for tens of millions of years. As a result, the anoxic sea floor became a

* Similar anoxic sea-floor conditions are found in certain areas today, such as the bottom of the Black Sea or the region of upwelling off the coast of Peru,[38] but during the Cretaceous these were widespread around the world.

thick sludge of organic-rich mud, which turned into extensive deposits of black shale rock. The period during which huge areas of shale accumulated in the Tethys Sea has therefore been called 'the Black Death'.[39]

Our planet has seen both earlier and more recent episodes of formation of crude oil and natural gas, but by far the most prolific were the organic-loaded black shales deposited around the continental shelves of the Tethys Ocean during the late Jurassic and mid Cretaceous periods. The Persian Gulf, the most abundant region for oil and gas today, as well as the substantial deposits in western Siberia, the Gulf of Mexico, the North Sea and Venezuela, were all produced by the combination of geological processes at this time.[40]

CUTTING OUT THE MIDDLEMAN

While coal powered the Industrial Revolution and oil carried us into our modern technological civilisation, humanity's exploitation of these fossil fuels has brought with it some now well-established global problems. Since the early seventeenth century we've been fervently digging up this buried ancient carbon that took tens of millions of years for the Earth to slowly stockpile, and we burned a great deal of it in just a few centuries. While there are concerns over peak oil and the diminishing supply of crude, there is plenty of accessible coal still underground – certainly another few centuries' worth at current consumption rates.[41] In this sense, then, we're not currently facing another energy crisis but a climate crisis, born as a result of our past solution to our energy hunger.

The carbon dioxide released by the combustion of fossil fuels has been rapidly increasing its level in the atmosphere, which is now already 45 per cent higher than prior to the Industrial Revolution. Indeed, the current rate of greenhouse gas emissions from human civilisation is unprecedented in the geological history of at least the last 66 million years. Perhaps the closest natural analogue was the Palaeocene–Eocene Thermal

Maximum[42] that we explored in Chapter 3 and which saw a rapid increase in global temperature making the world 5–8 °C hotter than today.[43] We're currently doing our best (or worst) to yank our global climate back to that period.[44]

The presence of such greenhouse gases in the atmosphere is not in itself a problem – indeed, it is their insulating effect that through our planet's history has kept the Earth's surface above freezing and so has been vital for supporting complex life.* But the rapidly rising carbon dioxide level is shifting the current established equilibriums in the natural world, and impacting on how we support our civilisation. It has caused increasingly acidic oceans, threatening coral reefs as well as the fisheries we rely upon for food.[45] Moreover, a warming global climate in turn drives rising sea levels that threaten our coastal cities, and shifts in the world's rainfall patterns have significant implications for agriculture.

But carbon dioxide is not the only form of pollution released by fossil fuels. As we saw, oxygen-poor conditions were required to prevent the decomposition of dead organisms and so allow the carbon to accumulate and become coal, oil and natural gas. These same conditions also favour the formation of sulphide compounds – this is why bogs today often have the distinctive rotten-egg smell of hydrogen sulphide – which are released when fossil fuel is burned and reacts with moisture in the air to create sulphuric acid. Thus the oxygen-poor soil of the Carboniferous coal swamps and the sediment of the Cretaceous sea floor also locked up future acid rain.[46]

Burning fossil fuels has been like releasing a trapped genie: it granted us our seventeenth-century wish for virtually limitless energy, but has done so with mischievous malice for the unintended consequences further down the line.

The challenge facing us now is to reverse the trend since the Industrial Revolution and once again decarbonise our economy. As we saw earlier in the chapter, throughout history, our

---

* We saw in Chapter 6 how the Great Oxidation Event created the iron ores that we have mined through history, but also scrubbed the atmosphere of the greenhouse gas methane to trigger a snowball Earth.

intensification of agriculture and harvesting of woodland has enhanced the rate at which humanity could gather solar energy. This sunlight is transformed into nutrition for our bodies, as well as into the raw materials and fuel we need, and we learned how to harness mechanical power from the natural world with waterwheels and windmills. Part of the solution to our current carbon crisis will be to return to these age-old practices, but with technological updates. Farms of solar panels produce electricity directly, and hydroelectric dams and wind turbines are identical in principle to waterwheels and mills, although prodigiously more productive than their technological forebears.

But perhaps the next revolution in humanity's enduring efforts to marshal ever greater supplies of energy will be to crack nuclear fusion: to harness the power source of the stars themselves. We saw in Chapter 6 how nuclear fusion within stars fuses hydrogen atoms together to create helium, and releases a great deal of energy in the process. Several facilities around the world are already making good progress in scaling up experimental reactors for mainstream nuclear power stations. The fusion fuel can be extracted from seawater, and the operation of such reactors produces no carbon dioxide or long-lived radioactive waste. So fusion offers not only abundant energy, but this time also cleanly. In this sense we will have come full circle: from the earliest agrarian societies capturing the energy of sunlight with their fields of crops and felled woodland, to installing a miniature sun within our fusion reactors, and so cutting out the middleman.*

---

* The atmospheric carbon dioxide levels won't naturally return to their pre-industrial setting for tens of thousands of years. The overlapping rhythms of the Milankovitch cycles are due to push the Earth's climate back to glaciation around 50,000 years from now, but the sharp shove we've already delivered to the atmosphere means that this next scheduled ice age will almost certainly be skipped. So from a human perspective, one silver lining to the current global warming might be that our civilisation will be better able to adapt in the long run to the extremes of a hotter world than the return of kilometre-thick ice sheets grinding across the Northern Hemisphere and a punishingly cold and dry climate making widespread agriculture impossible.[47]

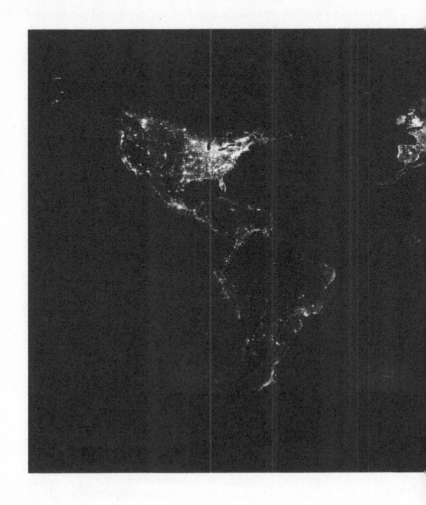

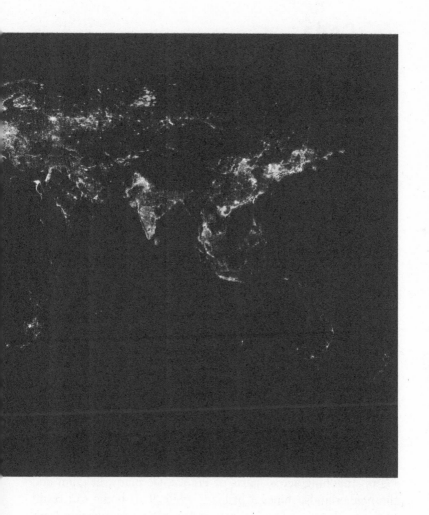

Coda

The human world is now clearly visible from space, highlighted by the electric brilliance of our towns and cities – a sparkling galaxy of artificial stars. This composite image was created by satellite, photographing the vista below on clear nights and then stitched together into a single omniscient view of the Earth from the heavens. In this way it's almost an abstraction, depicting the whole world simultaneously at night time and without any veiling of clouds. And it isn't a complete map of human habitation – much of the world's population in developing nations is still rural – but of industrialised urbanisation. Still, I think it beautifully illustrates the global civilisation we have built over the millennia, and how we've been moulded by the planet we live on.

The densest concentrations of humanity are immediately apparent: northern India and Pakistan, the Chinese plain and coastline – two of the earliest cradles of civilisation – as well as the lattice of cities and highways in the eastern US, grading gently into the central prairies. The crowded North European Plain, stretching across parts of France, Germany, Belgium and the Netherlands, shines a brilliant white. This is the end result of the gradual but decisive shift in population distribution from the Mediterranean rim to Northern Europe through the first millennium AD, driven by the use of iron-edged axe and plough that transformed the forests and damp clay soil into highly productive farmland. The intricate outline of the Mediterranean – the puddle remnant of the once-vast ancient Tethys sea – can be clearly made out, especially the bright coastal strip in the east showing up the crowded urbanisation of Israel, Lebanon and Syria.

Just as revealing are the dark areas on land. These are the landscapes and climate zones unsuited for dense human habitation. Mountain ranges are conspicuous by their invisibility: the shining furrow of the Po Valley at the top of Italy is capped by the gloomy Alps, the intense gleam of northern India abruptly cut by the curve of the Himalayas. Deserts appear as patches of broad darkness in the heart of Australia, southern Arabia and northern Africa. The ribbon oasis of the Nile Valley and its delta burns like a river of fire through this otherwise inhospitable region. The radiant triangle of the Indian subcontinent also stands out within the band of deserts reaching around the planet, dampened by the monsoons that seasonally suck in moisture from the enveloping ocean.

And it's not just the hyper-arid regions of the world that have hindered inhabitation, but also the equatorial zone of our planet with its high precipitation and therefore dense rainforest: Central Africa, the Amazon and the heart of Indonesia. These absences of electric light reveal both the rainy rising arm and the dry descending zone of the Hadley cell, the circulation current in the Earth's atmosphere.

Within Asia, the glittering froth of human activity is broken by the dark cavities of the freezing heights of the Tibetan plateau and the deserts of the continental interior. And running east–west through the heart of the continent are two roughly parallel stripes of diffuse glow. The more southern streak is the old course of the Silk Road, threaded between the mountains and deserts. It once carried commerce and knowledge across the breadth of Eurasia, connecting cultures on the continent's extremities, and today its imprint is still visible from space by the electric lustre of the cities that grew from the ancient oasis towns and entrepôts. The northern band follows the ecological zone of the grassy steppes, once an unknown wilderness from where nomadic pastoralists threatened the agrarian civilisations around the continental rim. The western half of this zone has now been put to the plough and transformed into great swaying fields of wheat, feeding the new cities all along this climate band, strung like pearls along the Trans-Siberian Railway.

You might think that other features of the planet that played such a pivotal role in our history ought not to be perceptible in this map of human light – such as the global pattern of contrary wind bands and the great swirling currents of the ocean gyres. We exploited them to build vast intercontinental trade networks and maritime empires, which in turn provided the raw materials and economic drivers for the Industrial Revolution. But although currents in the air and sea are invisible, their effects are still revealed in this image. The lights of fishing boats can be discerned from space, swarming like fireflies in the coastal regions where ocean upwelling brings nutrient-rich waters to the surface and where plankton – and the fish that feed on them – can thrive, such as along the continental shelf of Peru. And the glow of Norway, Sweden and Finland reveals habitation far further north than in the corresponding latitudes in Canada and Siberia. This is due to their milder climate granted by westerly winds blowing over the ocean and by the Gulf Stream – they are warmed by transported Caribbean sunshine. Even the deep subterranean reservoirs of fossil energy are rendered visible by flare stacks burning the natural gas released in the oilfields of the North Sea, the Persian Gulf and northern Siberia.

This single image encapsulates the culmination of our human story thus far – and we have come a long way since our origins. The Earth is a restlessly dynamic place, and its facial features and planetary processes have played a decisive role throughout the human story. Our species emerged within the unique tectonic and climatic conditions of the East African Rift, where the versatility and intelligence that allowed us to progress from apeman to spaceman were bestowed by environmental fluctuations driven by cosmic cycles. And before that, the intense temperature spike of the PETM 55.5 million years ago saw the emergence and rapid dispersal of our lineage, the primates, as well as the orders of ungulate mammals whose descendants we came to domesticate. Other global changes have been more gradual, such as the overall cooling and drying trend over the past few tens of millions of years that drove the spread of the grass species we came to cultivate as cereal crops. This planetary chilling culminated in the current period of flickering ice ages which moulded much of the landscape and allowed our species to populate the world.

The entire history of civilisation is just a flash in the current interglacial period – a transient spell of climatic stability. During these past few millennia we've dug up Earth's stony subterranean layers and piled them above ground to construct our buildings and monuments. We've excavated rich ores where metals have been concentrated by particular geological processes. And in the last few centuries we've mined the coal formed during a quirky period of the planet's past when ancient forests refused to rot, and we've sucked up the oil created by plankton settling to the asphyxiated seafloor of a drowned world.

We've now turned over a third of the Earth's total land area to agriculture. Our mining and quarrying moves more material than all the world's rivers combined.[1] And our industrial exhalations release more carbon dioxide than volcanoes, warming the climate of the entire planet. We have profoundly altered the world, but we only recently acquired such overwhelming dominion over Nature. The Earth set the stage for the human story and its landscapes and resources continue to direct human civilisation.

The Earth shaped our history.

website

Explore further material, reading recommendations, and videos on the book's website:

www.originsbook.com

🐦 @lewis_dartnell

🐦 @OriginsBook

NOTES

INTRODUCTION

1. For a much more complete description of how the elements in the human body came from the Earth see Stager (2014) and Schrijver (2015). • 2. Crutzen (2000); Ruddiman (2015); Lewis (2015). • 3. Dartnell (2015).

1 THE MAKING OF US

1. Arnason (1998); Patterson (2006); Moorjani (2016). • 2. Rothery (2010), p. 53. • 3. Cane (2001). • 4. King (2006). • 5. Stow (2010), Kindle location 740. • 6. Maslin (2014). • 7. Ibid. • 8. Jung (2016). • 9. Maslin (2013); Shubin (2014), p. 179; Fer (2017). • 10. Cane (2001). • 11. Lieberman (2014), p. 68. • 12. Chorowicz (2005). • 13. King (1994). • 14. King (2006); Bailey (2011). • 15. Maslin (2014). • 16. Ibid. • 17. Berna (2012). • 18. Gibbons (1998). • 19. Ermini (2015). • 20. Bramble (2004). • 21. Bradley (2008). • 22. Maslin (2014). • 23. White (2003). • 24. Potts (2013). • 25. Maslin (2007). • 26. Maslin (2007); Trauth (2010). • 27. Maslin (2007). • 28. Maslin (2007); Trauth (2010). • 29. Trauth (2010). • 30. Maslin (2014); Potts (2015). • 31. Trauth (2007); Maslin (2007). • 32. Maslin (2007). • 33. Potts (2015). • 34. Maslin (2014). • 35. Ibid. • 36. Neimark (2012). • 37. Ibid.; McKie (2013). • 38. Jung (2016). • 39. Giosan (2012). • 40. Reilinger (2011). • 41. Garzanti (2016). • 42. US Geological Survey publications, 'Plate tectonics and people', https://pubs.usgs.gov/gip/dynamic/tectonics.html. • 43. Shuckburgh (2008), p. 133. • 44. This section on the link between ancient civilisations and plate boundaries: Force (2008); Force (2010); Force (2012); Force (2015), Ch. 15. • 45. Jackson (2006). • 46. http://

worldpopulationreview.com/world-cities/tehran-population. • 47. Jackson (2006); Shuckburgh (2008), p. 133.

2 CONTINENTAL DRIFTERS

1. Kukula (2016), Kindle location 4136; Ruddiman (2016), Ch. 4. • 2. Woodward (2014), p. 28. • 3. Ibid., p. 111. • 4. Stager (2012), Kindle location 305. • 5. Ibid. • 6. Summerhayes (2015), p. 264. • 7. Ibid. • 8. Ruddiman (2016), p. 45. • 9. Feurdean (2013); Liddy (2016). • 10. Summerhayes (2015), p. 255. • 11. Franks (1960). • 12. Woodward (2014), p. 102. • 13. Ibid., p. 112. • 14. Ibid.. • 15. Ibid., p. 116. • 16. Maslin (2014). • 17. Ruddiman (2016), p. 42. • 18. Lenton (2013), p.353; Woodward (2014), p. 111. • 19. Nield(2014), p.213; Woodward (2014), p. 35. • 20. Ibid. • 21. Stow (2010), p. 131. • 22. Summerhayes (2015), p. 368; Stager (2012), Kindle location 1178. • 23. Woodward (2014), p. 121; Lieberman (2014), p.68; Ruddiman (2016), p. 42. • 24. Woodward (2014), p. 121. • 25. Woodward (2014), p. 121; Maslin (2007). • 26. Mendez (2011). • 27. Woodward (2014), p. 121; O'Dea (2016). • 28. Maslin (2007); Woodward (2014), p.121; Ruddiman (2016), p. 19. • 29. Summerhayes (2015), p. 369. • 30. Oppenheimer (2011), p. 176. • 31. Bowden (2012). • 32. Ermini (2015); Lopez (2015); Tucci (2016). • 33. Eriksson (2012); Lenton (2013), p. 367. • 34. Oppenheimer (2011), p. 178; King (2006). • 35. Morris (2011), Kindle location 1274; Lieberman (2014), p. 130. • 36. Ermini (2015). • 37. Abi-Rached (2011). • 38. Ermini (2015). • 39. Carotenuto (2016). • 40. Eriksson (2012). • 41. Lenton (2013), p. 367. • 42. Oppenheimer (2011), p. 179. • 43. Eriksson (2012). • 44. Ibid. • 45. Paine (2013), p. 14. • 46. McNeill (2012). • 47. Woodward (2014), p. 29. • 48. Ibid. • 49. Morris (2011), Kindle location 1444. • 50. Holen (2017). • 51. Rose (2011). • 52. Bradley (2008). • 53. McNeill (2012). • 54. Ermini (2015). • 55. McNeill (2012). • 56. US Geological Survey publications, 'Past Glaciations and '"Little Ice Ages"', https://pubs.usgs.gov/pp/p1386i/chile-arg/wet/past.html. • 57. Novacek (2008), p. 267. • 58. Discussion on the consequences for American history of a less icy Ice Age appeared in Dutch (2006); Alvarez (2018), p. 68. • 59. Stager (2012), Kindle location 305; Summerhayes (2015), p. 264. • 60. Woodward (2014),

p. 29. • 61. Ruddiman (2016), p. 44. • 62. Gibbard (2007). • 63.
Gibbard (2007); Gupta (2007); Gupta (2017). Discussion on the
consequences for European history of the last Ice Age appeared in
Dutch (2006). • 64. Frankopan (2016), p. 387. • 65. Kaplan (2017),
Kindle location 643. • 66. Marshall (2016), p. 91; Kaplan (2017),
Kindle location 650. • 67. Frankopan (2016), p. 386.

3 OUR BIOLOGICAL BOUNTY

1. Shakun (2012). • 2. Murton (2010). • 3. Törnqvist (2012);
Summerhayes (2015), p. 255. • 4. McNeill (2012). • 5. Belfer-Cohen
(1991); Bar-Yosef (1998); Grosman (2008). • 6. Teller (2002); Tarasov
(2005); Woodward (2014), p. 130. • 7. Belfer-Cohen (1991); Bar-Yosef
(1998); J. R. McNeill (2004), p. 23; Grosman(2008); Balter (2010);
Shubin (2014), p. 177. • 8. McBrearty (2000); Sterelny (2011). • 9.
Ruddiman (2016), p. 63. • 10. McNeill (2012); Lenton (2013), p.
369. • 11. White (2012); Balter (2010). • 12. Morton (2016), p.
226. • 13. Richerson (2001). • 14. Hodell (1995); Mayewski (2004).
And see, for general discussion, Diamond (2011); Cowie (2012); Brooke
(2014). • 15. McNeill (2012) • 16. Petit (1999); Wright (2006), p.
50. • 17. Sage (1995); Richerson (2001); Morton (2016), p. 229. • 18.
Kilian (2010). • 19. Lenton (2013), p. 369. • 20. Ruddiman (2016),
p. 71. • 21. McNeill (2004), p. 33. • 22. Ruddiman (2016), p. 70. • 23.
McNeill (2004), p. 32. • 24. Londo (2006); Ruddiman (2016), p.
70. • 25. Kimber (2000). • 26. Larson (2014). • 27. Ibid. • 28. Ibid.
For general references on the locations and timings of domestication
of crops see Diamond (2003); Fuller (2014); Larson (2014). • 29.
Janzen (1982); Guimaraes (2008); Yong (2015). • 30. Larson
(2014). • 31. Thompson (2010), p. 27; McNeill (2012). • 32. Lieberman
(2014), p.188; Marr (2013), Kindle location 563. • 33. Marr (2013),
Kindle location 563. • 34. Lenton (2013), p. 368. • 35. Brooks
(2006); Morris (2011), Kindle location 2867. • 36. Reader
(2005), p. 27. • 37. McNeill (2004), p. 43. • 38. Reader (2005),
p. 25. • 39. Stager (2012), Kindle location 3153. • 40. Brooks
(2005). • 41. Brooks (2006); de Blij (2011), p. 142;
Nicoll (2013). • 42. Allen (1997); Morris (2011), Kindle location
2945; White (2012); Nicoll (2013). • 43. Brooks (2006); White
(2012). • 44. Marr (2013), p. 64. • 45. Wright (2006), p. 102;

Marshall (2016), p. 108. • 46. Marshall (2016), p. 108. • 47. McNeill (2004), p. 53.• 48. Ibid., p. 43. • 49. Ermini (2015). • 50. McNeill (2004), p. 29.• 51. Larson (2014). • 52. Ibid. • 53. Anthony (2010), p. 102. • 54. McNeill (2004), p. 31. • 55. Curry (2013). • 56. McNeill (2004), p. 31. • 57. Ibid. • 58. Anthony (2010), Kindle location 2568. • 59. Watson (2012), p. 139. • 60. International Energy Agency, https://www.iea.org/topics/coal. • 61. Thompson (2010), p. 70; Hanson (2016), p. 67. • 62. Novacek (2008), p. 153. • 63. Stow (2010), p. 146. • 64. Kourmpetli (2014). • 65. Kellogg (2001); Novacek (2008), p. 226. • 66. Hanson (2016), p. 75. • 67. Lenton (2013), p. 340. • 68. Bryan (2015), p. 136. • 69. Quran 2:173. Sahih International translation, quran.com. • 70. Gingerich (2006). • 71. Wing (2005); McInerney (2011). • 72. Nield (2014), p. 211. • 73. Weijers (2007). • 74. McInerney (2011). • 75. Woodburne (2009). • 76. Gingerich (2006). • 77. Gingerich (2006); McInerney (2011); Gehler (2016). • 78. McInerney (2011). • 79. Bowen (2002); Gingerich (2006); McInerney (2011). • 80. Diamond (1998), p. 140; Morris (2011), Kindle location 1979. • 81. Diamond (1998), Ch. 10; see also McNeill (2001); Ramachandran (2011); Laitin (2012). • 82. Bernstein (2009), p. 70. • 83. Diamond (1998), pp. 159–62. • 84. Stahl (2008). • 85. Bernstein (2009), p. 19. • 86. Diamond (1998). • 87. Twinning (2009); Marshall (2016), p. 38. • 88. https://www.worldwildlife.org/stories/the-earth-has-a-third-pole-and-millions-of-people-use-its-water; Sinha (2010); Qiu (2014). • 89. Stow (2010), p. 188; Qiu (2014). • 90. https://www.worldwildlife.org/stories/the-earth-has-a-third-pole-and-millions-of-people-use-its-water; Sinha (2010).• 91. Kaplan (2017), p. 225. • 92. Sinha (2010). • 93. Ibid. • 94. Lim (2004); Wong (2010).

4 THE GEOGRAPHY OF THE SEAS

1. Stewart (1994); Hoffecker (2005); Hu (2009). • 2. Rick (2008), p. 230; Barrett (2011); Sahrhage (2012), Ch. 2. • 3. Fagan (2001), Kindle location 879; Pye (2015), p. 177. • 4. Fagan (2001), Kindle location 215–60, 820, 910; Hoffman (2017), p. 115. • 5. Pye (2015), p. 259. • 6. Bernstein (2009), p. 272. • 7. Marr (2013), p. 356. • 8. Bernstein (2009), p. 273. • 9. Allen (2009), p. 138. • 10. Henrich

(2004). • 11. Leidwanger (2014). • 12. Force (2015), p. 143. • 13. Fernandez-Armesto (2002), p. 361. • 14. Véron (2006). • 15. Brotton (2013), p. 17. • 16. Maslin (2007); Garcia-Castellanos (2009); Stow (2010), Kindle location 470. • 17. Woodward (2014), p. 121. • 18. Stow (2010), Kindle location 472; Krijgsman (1999). • 19. Maslin (2007). • 20. Garcia-Castellanos (2009). • 21. Paine (2013), p. 3; Bernstein (2009), p. 44. • 22. Stavridis (2018), p. 23. • 23. East (1967), p. 170. • 24. Bernstein (2009), p. 18. • 25. East (1967), p. 170; Bernstein (2009), p. 50. • 26. Bernstein (2009), p. 18. • 27. Bernstein (2009), p. 94; Braudel (1995), p. 55. • 28. Bernstein (2009), p. 145. • 29. Oppenheimer (2011), Ch. 8.4; Gatti (2012). • 30. Paine (2013), p. 168. • 31. Crowley (2016), Kindle location 852. • 32. Frankopan (2016), p. 329. • 33. Paine (2013), p. 281. • 34. Hanson (2016), p. 131. • 35. Bernstein (2009), p. 52. • 36. Bernstein (2009), p. 142; Paine (2013), p. 280. • 37. Bernstein (2009), p. 142; Paine (2013), p. 280. • 38. Paine (2013), p. 281; Frankopan (2016), p. 271. • 39. Bernstein (2009), p. 134. • 40. Paine (2013), p. 169. • 41. Bernstein (2009), p. 134; Crowley (2016), Kindle location 3865. • 42. Crowley (2016), Kindle location 3865. • 43. Bernstein (2009), p. 141. • 44. McNeill (1963), p. 194; Marr (2013), p. 94. • 45. McNeill (1963), p. 198; Marr (2013), p. 95; Kaplan (2017), Kindle location 865. • 46. Reader (2005), p. 53. • 47. Bernstein (2009), p. 57. • 48. Fromkin (2000), p. 70. • 49. Bernstein (2009), p. 454. • 50. World Oil Transit Chokepoints, https://www.eia.gov/beta/international/regions-topics.cfm?RegionTopicID=WOTC. • 51. Ibid. • 52. http://news.bbc.co.uk/onthisday/hi/dates/stories/november/29/newsid_3247000/3247805.stm; McDermott (1998), pp. 136, 142. • 53. Corones (2015). • 54. Friedman (2017). • 55. World Oil Transit Chokepoints, https://www.eia.gov/beta/international/regions-topics.cfm?RegionTopicID=WOTC. • 56. Stern (2010). • 57. Marshall (2016), p. 143. • 58. Friedman (2017). • 59. This argument on the link between voting patterns and an ancient sea is from Dutch (2002). • 60. US election 2016: Trump victory in maps, http://www.bbc.co.uk/news/election-us-2016–37889032.

5 WHAT WE BUILD WITH

1. Haywood (2012), p. 12; Morris (2011), Kindle location 997. • 2. Details on the construction of the Great Pyramid in Verner (2001), Ch. 3; Sweeney (2007), p. 16. • 3. Stow (2010), p. 166. • 4. Ibid. • 5. Fortey (2010), p. 284. • 6. Bernstein (2009), p. 39. • 7. Phillips (1988). • 8. Ibid. • 9. The Getty Centre: Architecture, http://www. getty.edu/visit/center/architecture.html. • 10. Siddall (2015). • 11. Nield (2014), p. 47. • 12. Brison (2005); Sakellariou (2016), p. 168. • 13. Stow (2010), p. 135. • 14. Pollard (2017). • 15. Sheridan (2002). • 16. Stow (2010), p. 133; Stampfli (2013). • 17. Rasmussen (2012), p. 45. • 18. Chen (2012). • 19. Wignall (2017), p. 64. • 20. Ibid., p. 9. • 21. Ibid., p. 161. • 22. Ibid. • 23. Ibid., p. 169. • 24. Myers (1997); Fortey (2005), p. 304. • 25. Fortey (2005), p. 297. • 26. Koestler-Grack (2010), p. 39. • 27. Nield (2014), p. 140. • 28. Ibid. • 29. Kneller (1987). • 30. Zalasiewicz (2012), p. 42. • 31. Fortey (2010), p. 171. • 32. Fortey (2005), p. 309. • 33. Gregory (2010), p. 22. • 34. Fortey (2010), p. 248. • 35. Ibid., p. 97. • 36. Schuberth (1968), p. 81; Barr (2011). • 37. Fortey (2005), p. 243; Barr (2011). • 38. Winkless (2017).

6 OUR METALLIC WORLD

1. De Ryck (2005); Roberts (2009). • 2. Bernstein (2009), p. 37. • 3. Goody (2012), p. 9. • 4. Bernstein (2009), p. 37. • 5. Fokkens (2013), p. 420. • 6. Fortey (2005), p. 294. • 7. Bernstein (2009), p. 38. • 8. Kassianidou (2013). • 9. Candela (2005), p. 423. • 10. Fortey (2005), p. 188. • 11. Ibid. • 12. Republic of Cyprus: Geological Survey Department, http://www.moa.gov.cy/moa/gsd/gsd.nsf/ dmltroodos_en/dmltroodos_en. • 13. Cann (2004). • 14. Stow (2010), p. 200. • 15. Kassianidou (2013). • 16. Ibid. • 17. Constantinou (1982); Socratous (2011); Kassianidou (2013). • 18. Wagner (2009), p. 98. • 19. Hughey (2013). • 20. Marr (2013), p. 70. • 21. Ibid.. • 22. Angelakis (2006); Oppenheimer (2011), p. 279. • 23. Oppenheimer (2011), p. 279. • 24. Ibid., p. 233. • 25. Oppenheimer (2011), p. 278. • 26. Ibid., pp. 278, 293. • 27. Ibid., p. 292. • 28. Winchester (2011), pp. 62–8. • 29. Stow (2010), p. 203. • 30. Roebroeks (2012). • 31. Osborne (2013), Needham

(1965), p. 370. • 32. Mokyr (1992), p. 210. • 33. Oleson (2009), p. 170. • 34. McNeill (2004), pp. 101–102. • 35. Andersen (2016). • 36. Hillstrom (2005), p. 11. • 37. Kasen (2017). • 38. Sagan (1973), p. 190; Shubin (2014), p. 33; Schrijver (2015), p. 129; Kukula (2016), Kindle location 236. • 39. Kleine (2011). • 40. Walker (2006). • 41. Ridley (2013), p. 297. • 42. Klein (2005); Shubin (2014), p. 81. • 43. Lenton (2013), p. 243. • 44. Hamilton (2016). • 45. Lenton (2013), p. 183. • 46. Lyons (2014). • 47. Lenton (2013), pp. 29–33; Lyons (2014); Falkowski (2015), p. 88. • 48. Lenton (2013), p. 30. • 49. Scott (2006). • 50. Hagelüken (2014). • 51. Rohrig (2015). • 52. Graedel (2014). • 53. Schwarz-Schampera (2014). • 54. Graedel (2014). • 55. Belli (2007). • 56. Gunn (2014). • 57. Ibid. • 58. Sadykov (2000). • 59. Stewart (2005); Erisman (2008). • 60. Krivolutskaya (2016). • 61. Gunn (2014). • 62. Ibid. • 63. Humphreys (2014). • 64. Gunn (2014). • 65. Kinnaird (2005); Ridley (2013), p. 61. • 66. American Chemical Society: Endangered Elements, https://www.acs.org/content/acs/en/greenchemistry/research-innovation/research-topics/endangered-elements.html. • 67. Graedel (2014); Humphreys (2014). • 68. Clapper (2013), p. 11. • 69. Jacobs (2018). • 70. Lewis (2008); Warren (2014); Wagland (2016). • 71. Kolarik (2005).

7 SILK ROADS AND STEPPE PEOPLES

1. McNeill (2014), p. 65. • 2. Kaplan (2017), p. 212. • 3. Marshall (2016), p. 31. • 4. Kaplan (2017), p. 212. • 5. Pye (1995); Vasiljevic (2014). • 6. Wright (2006), p. 104. • 7. Haase (2007); Vasiljević (2014). • 8. Fromkin (2000), p. 41. • 9. Braudel (1995), p. 352; Wel (2014). • 10. East (1967), p. 166; Marr (2013), p. 139. • 11. Millward (2013), p. 78. • 12. Ibid., p. 81. • 13. East (1967), p. 168; Bernstein (2009), p. 9. • 14. Bernstein (2009), p. 9. • 15. Ibid. • 16. Millward (2013), p. 80. • 17. Watson (2012), p. 462. • 18. East (1967), p. 175. • 19. Ji (2009). • 20. Orlando (2016); Marr (2013), p. 253. • 21. Piantadosi (2003), p. 82. • 22. Ibid., p. 81. • 23. McNeill (2004), p. 95. • 24. Bernstein (2009), p. 74. • 25. McNeill (2004), p. 98. • 26. McDougall (1983); McDougall (1990); Marr (2013), p. 254. • 27. Frankopan (2016), p. 30. • 28. Braudel (1995), p. 63. • 29. East (1967), p. 175. • 30. Bernstein (2009),

p. 139. • 31. Millward (2013), p. 22. • 32. Barry (2014), p. 146. • 33. Millward (2013), p. 23. • 34. Marr (2013), p. 337. • 35. Millward (2013), p. 24. • 36. Marr (2013), p. 337. • 37. Millward (2013), p. 60. • 38. Anthony (2010), Kindle location 3521. • 39. Millward (2013), p. 95. • 40. Anthony (2010), p. 101. • 41. Anthony (2010), Kindle location 6495; Millward (2013), p. 95. • 42. Keegan (1993), p. 240. • 43. Watson (2012), p.288. • 44. Schmidt (2017). • 45. Fortey (2005), p. 471. • 46. Bernstein (2009), p. 113. • 47. Millward (2013), Kindle location 350. • 48. McNeill (2004), pp. 100–101. • 49. Fernandez-Armesto (2002), p. 115. • 50. Keegan (1993), p. 190; Kaplan (2017), Kindle location 1196. • 51. Keegan (1993), pp. 180, 206. • 52. Marshall (2016), p. 33. • 53. East (1967), p. 66. • 54. Ibid., p. 68. • 55. Keegan (1993), p. 183. • 56. Ibid., pp. 184, 186; McNeill (2004), p. 100. • 57. Keegan (1993), p. 212. • 58. Frankopan (2016), p. 289. • 59. Millward (2013), p. 41; Marr (2013), p. 262. • 60. Frankopan (2016), p. 75. • 61. McCormick (2012), p. 190; Brooke (2014), p. 347. • 62. McCormick(2012), p. 190. • 63. Keegan (1993), pp. 184, 186; Fromkin (2000), p. 97. • 64. Keegan (1993), p. 187. • 65. Frankopan (2016), p. 76. • 66. Ibid. • 67. Ibid., p. 77. • 68. Marr (2013), p. 264. • 69. Millward (2013), p. 237. • 70. Ibid. • 71. Ibid., p. 47; Frankopan (2016), p. 237. • 72. Frankopan (2016), p. 237. • 73. Millward (2013), p. 47. • 74. Frankopan (2016), p. 242. • 75. Ibid., p. 244. • 76. Marr (2013), p. 264. • 77. Frankopan (2016), p. 244. • 78. East (1967), p. 177. • 79. Frankopan (2016), p. 249. • 80. Marr (2013), p. 264; Frankopan (2016), p. 249. • 81. Keegan (1993), p. 201; Marr (2013), p. 264. • 82. Millward (2013), p. 48; Frankopan (2016), p. 240. • 83. Frankopan (2016), p. 241. • 84. Ibid., p. 261. • 85. Bernstein (2009), pp. 113, 146; Watson (2012), p. 461; Millward (2013), p. 48. • 86. McNeill (2004), p. 124 Millward (2013), p. 48. • 87. Watson (2012), p. 462. • 88. Fernandez-Armesto (2002), p. 131. • 89. Frankopan (2016), p. 274. • 90. Fagan (2001), Kindle location 598, 949; McNeill (2004), p. 120. • 91. Marr (2013), p. 278. • 92. Frankopan (2016), p. 278. • 93. Marr (2013), p. 279. • 94. Morris (2011), p. 390; Marr (2013), p. 265. • 95. Bernstein (2009), p. 160. • 96. Roberts (1967); Parker (1976); Rogers (2018); Morris (2011), Kindle location 7186. • 97. McNeill (2004), p. 194. • 98. Ibid., p. 195. • 99. Morris (2011), Kindle location 7186. • 100. Millward (2013), Kindle location 1780. • 101. Vasiljevic (2014). • 102. Frankopan (2016), pp. 524, 718. • 103. Price (2009); Smith (2009). • 104. Watson (2012), p. 237. • 105. Frankopan (2016), p. 524 • 106. Terazono (2016).

8 THE GLOBAL WIND MACHINE AND THE AGE OF DISCOVERY

1. Fromkin (2000), p. 114; Crowley (2016), Kindle location 91–112. • 2. Ibid., Kindle location 190, 243; Paine (2013), p. 389. • 3. Rosenbaum (2002); Alvarez (2008), p. 73. • 4. Paine (2013), p. 377. • 5. Tomczak (1994), p. 422. • 6. Pim (2008); Ramalho (2010); The Geology of the Canary Islands, http://www.island-sinocean.com/view/The_Geology_of_Canary_Islands. • 7. Cannat (1999). • 8. Frankopan (2016), p. 300. • 9. Ibid., p. 298. • 10. Morris (2011), Kindle location 6535. • 11. Paine (2013), p. 385; Frankopan (2016), p. 298. • 12. Chamberlin (2013), p. 85; Paine (2013), p. 385. • 13. Winchester (2011), pp. 108–109. • 14. Paine (2013), p. 389. • 15. Chamberlin (2013), p. 72. • 16. Raudzens (2003), p. 216; Paine (2013), p. 386. • 17. Fromkin (2000), p. 116. • 18. Crowley (2016), Kindle location 411. • 19. Ibid., Kindle location 454. • 20. Ibid., Kindle location 248. • 21. Ibid., Kindle location 473. • 22. Paine (2013), p. 391. • 23. Bernstein (2009), p. 201; Marr (2013), p. 306; Paine (2013), p. 392. • 24. Paine (2013), pp. 64, 388; Bernstein (2009), p. 204. • 25. Chamberlin (2013), p. 85; Crowley (2016), Kindle location 264. • 26. Paine (2013), p. 396. • 27. Rodger (2012). • 28. Huang (2014). • 29. Winchester (2011), p. 116. • 30. Oxford English Dictionary, 2nd ed. (1989), Oxford University Press. • 31. Bernstein (2009), p. 209; Crowley (2016), Kindle location 732. • 32. Bernstein (2009), p. 214; Frankopan (2016), p. 321. • 33. Sauberlich (1997). • 34. Crowley (2016), Kindle location 1296. • 35. Clift (2008); Boos (2010); Raj (2013); Rajagopalan (2013). • 36. Crowley (2016), Kindle location 844, 852. • 37. Ibid., Kindle location 1374. • 38. Crowley (2016), p. 336. • 39. Brotton (2013), p. 189. • 40. Paine (2013), p. 376. • 41. Rodger (2012). • 42. Brotton (2013), p. 191. • 43. Ibid., p. 196. • 44. Ibid., p. 211. • 45. Ibid., p. 198. • 46. Bernstein (2009), pp. 247–8. • 47. Fish (2010), p. 360; Headrick (2010), p. 40. • 48. McNeill (2004), p. 202; Bernstein (2009), p. 249; Paine (2013), p. 407; Frankopan (2016), p. 335. • 49. Cunningham (1996); Waltham (2005). • 50. Cunningham (1996). • 51. Frankopan (2016), p. 335. • 52. Braudel (1995), p. 444. • 53. Bernstein (2009), p. 247; Paine (2013), p. 404; Frankopan (2016), p. 335. • 54. McNeill (2004), p. 202; Morris (2011), Kindle location 439. • 55. Frankopan (2016),

p. 341. • 56. Braudel (1995), p. 444; Waltham (2005). • 57. Bernstein (2009), p. 259. • 58. Ibid., p. 297. • 59. McNeill (2004), p. 169; Paine (2013), p. 410. • 60. Marr (2013), p. 441. • 61. Jones (2004), p. 41. • 62. Bernstein (2009), p. 341; Morris (2011), Kindle location 7290. • 63. Bernstein (2009), p. 338. • 64. McNeill (2004), p. 169. • 65. Marr (2013), p. 442.

9 ENERGY

1. Rackham (2009); Monbiot (2014) pp. 91–2. • 2. Morris (2011), Kindle location 7721. • 3. Needham (1965), p. 370. • 4. Leveau (1996); Morris (2011), Kindle location 4617. • 5. Greene (2000). • 6. Bithas (2016). • 7. Marr (2013), p. 272. • 8. Smith (1997). • 9. Bithas (2016). • 10. Kenrick (1997); Karol (2001). • 11. Nelsen (2016). • 12. Hanson (2016), p. 56. • 13. Lenton (2013), p. 307. • 14. Hanson (2016), p. 56. • 15. Weng (2010). • 16. Nelsen (2016). • 17. Ibid. • 18. Lenton (2013), p. 307. • 19. Fortey (2010), p. 168. • 20. Thomas (2013), p. 53; Shubin (2014), p. 82; Wignall (2017), p. 171. • 21. Morris (2011), Kindle location 470. • 22. McNeill (2004), p.231. • 23. Adams (2008); Schobert (2014), p. 64. • 24. Allen (2009), p. 235. • 25. Hillstrom (2005), p. 16. • 26. Moylan (2017). • 27. Macalister (2015). • 28. BP (2017). • 29. Speight (2015), p. 64. • 30. Dalvi (2015), p. 5. • 31. Bithas (2016), p. 8. • 32. Castree (2009), p. 273. • 33. Browne (2014), Kindle location 4203. • 34. Castree (2009), p. 273. • 35. Ulmishek (1990); Larson (1991). • 36. Zalasiewicz (2012), p. 165. • 37. Stow (2010), p. 131. • 38. Helly (2004). • 39. Stow (2010), p. 102. • 40. Stoneley (1990); Ulmishek (1990); Larson (1991); Mann (2003); Sorkhabi (2016). • 41. Lenton (2013), p. 54. • 42. Wright (2013); Zeebe (2016). • 43. McInerney (2011). • 44. Gingerich (2006); McInerney (2011); Stager (2012), Kindle location 1178; Wing (2013). • 45. Guinotte (2008); (Kroeker, 2010). • 46. Fortey (2010), p. 164. • 47. Stager (2012), Kindle location 489.

CODA

1. Douglas (2000); Zalasiewicz (2011); Zalasiewicz (2014).

Bibliography

Abi-Rached, L., M. J. Jobin, S. Kulkarni, A. McWhinnie, K. Dalva, L. Gragert, F. Babrzadeh, B. Gharizadeh, M. Luo and F. A. Plummer (2011). 'The shaping of modern human immune systems by multi-regional admixture with archaic humans', *Science*: 1209202.

Adams, S. P. (2008). 'Warming the Poor and Growing Consumers: Fuel Philanthropy in the Early Republic's Urban North', *Journal of American History* 95(1): 69–94.

Allen, R. C. (1997). 'Agriculture and the origins of the state in ancient Egypt', *Explorations in Economic History* 34(2): 135–54.

Allen, R. C. (2009). *The British Industrial Revolution in Global Perspective*, Cambridge University Press.

Alvarez, W. (2018). *A Most Improbable Journey: A Big History of Our Planet and Ourselves*, W. W. Norton & Co.

Andersen, T. B., P. S. Jensen and C. V. Skovsgaard (2016). 'The heavy plow and the agricultural revolution in Medieval Europe', *Journal of Development Economics* 118: 133–49.

Angelakis, A. N., Y. M. Savvakis and G. Charalampakis (2006). *Minoan Aqueducts: A Pioneering Technology*, International Water Association 1st International Symposium on Water and Wastewater Technologies in Ancient Civilizations, Iraklio, Greece.

Anthony, D. W. (2010). *The Horse, the Wheel, and Language: How Bronze-Age Riders from the Eurasian Steppes Shaped the Modern World*, Princeton University Press.

Arnason, U., A. Gullberg and A. Janke (1998). 'Molecular timing of primate divergences as estimated by two nonprimate calibration points', *Journal of Molecular Evolution* 47(6): 718–27.

Bailey, G. N., S. C. Reynolds and G. C. P. King (2011). 'Landscapes of human evolution: models and methods of tectonic geomorphology and the reconstruction of hominin landscapes', *Journal of Human Evolution* 60(3): 257–80.

Balter, M. (2010). 'The Tangled Roots of Agriculture', *Science* 327(5964): 404–406.

Bar-Yosef, O. (1998). 'The Natufian culture in the Levant, threshold to the origins of agriculture', *Evolutionary Anthropology* 6(5): 159–77.

Barr, J., T. Tassier and R. Trendafilov (2011). 'Depth to Bedrock and the Formation of the Manhattan Skyline, 1890–1915', *Journal of Economic History* 71(4): 1060–77.

Barrett, J. H., D. Orton, C. Johnstone, J. Harland, W. Van Neer, A. Ervynck, C. Roberts, A. Locker, C. Amundsen, I. B. Enghoff, S. Hamilton-Dyer, D. Heinrich, A. K. Hufthammer, A. K. G. Jones, L. Jonsson, D. Makowiecki, P. Pope, T. C. O'Connell, T. de Roo and M. Richards (2011). 'Interpreting the expansion of sea fishing in medieval Europe using stable isotope analysis of archaeological cod bones', *Journal of Archaeological Science* 38(7): 1516–24.

Barry, R. G. and E. A. Hall-McKim (2014). *Essentials of the Earth's Climate System*, Cambridge University Press.

Belfer-Cohen, A. (1991). 'The Natufian in the Levant', *Annual Review of Anthropology* 20: 167–86.

Belli, P., R. Bernabei, F. Cappella, R. Cerulli, C. J. Dai, F. A. Danevich, A. d'Angelo, A. Incicchitti, V. V. Kobychev, S. S. Nagorny, S. Nisi, F. Nozzoli, D. Prosperi, V. I. Tretyak and S. S. Yurchenko (2007). 'Search for α decay of natural Europium', *Nuclear Physics A* 789(1): 15–29.

Berna, F., P. Goldberg, L. K. Horwitz, J. Brink, S. Holt, M. Bamford and M. Chazan (2012). 'Microstratigraphic evidence of in situ fire in the Acheulean strata of Wonderwerk Cave, Northern Cape province, South Africa', *Proceedings of the National Academy of Sciences of the United States of America* 109(20): E1215–20.

Bernstein, W. L. (2009). *A Splendid Exchange: How Trade Shaped the World*, Atlantic Books.

Bithas, K. and P. Kalimeris (2016). 'A Brief History of Energy Use in Human Societies', *Revisiting the Energy-Development Link: Evidence from the 20th Century for Knowledge-based and Developing Economies*, ed. K. Bithas and P. Kalimeris, Springer International Publishing: 5–10.

Blij, H. d. (2011). *The Power of Place: Geography, Destiny, and Globalization's Rough Landscape*, Oxford University Press.

Boos, W. R. and Z. Kuang (2010). 'Dominant control of the South Asian monsoon by orographic insulation versus plateau heating', *Nature* 463: 218–23.

Bowden, R., T. S. MacFie, S. Myers, G. Hellenthal, E. Nerrienet, R. E. Bontrop, C. Freeman, P. Donnelly and N. I. Mundy (2012). 'Genomic Tools for Evolution and Conservation in the Chimpanzee: Pan troglodytes ellioti Is a Genetically Distinct Population', *PLoS Genetics* 8(3): 1–10.

Bowen, G. J., W. C. Clyde, P. L. Koch, S. Ting, J. Alroy, T. Tsubamoto, Y. Wang and Y. Wang (2002). 'Mammalian Dispersal at the Paleocene/Eocene Boundary', *Science* 295(5562): 2062–5.

BP (2017). BP Statistical Review of World Energy, June 2017.

Bradley, B. J. (2008). 'Reconstructing phylogenies and phenotypes: a molecular view of human evolution', *Journal of Anatomy* 212(4): 337–53.

Bramble, D. M. and D. E. Lieberman (2004). 'Endurance running and the evolution of Homo', *Nature* 432(7015): 345–52.

Braudel, F. (1995). *A History of Civilizations*, Penguin.

Brison, D. N. (2005). *Caves in the Odyssey*, 14th International Congress of Speleology. Kalamos, Hellas, Hellenic Speleological Society.

Brooke, J. L. (2014). *Climate Change and the Course of Global History*, Cambridge University Press.

Brooks, N. (2006). 'Cultural responses to aridity in the Middle Holocene and increased social complexity', *Quaternary International* 151: 29–49.

Brooks, N., I. Chiapello, S. D. Lernia, N. Drake, M. Legrand, C. Moulin and J. Prospero (2005). 'The climate-environment-society nexus in the Sahara from prehistoric times to the present day', *Journal of North African Studies* 10(3–4): 253–92.

Brotton, J. (2013). *A History of the World in Twelve Maps*, Penguin.

Browne, J. (2014). *Seven Elements that Have Changed the World: Iron, Carbon, Gold, Silver, Uranium, Titanium, Silicon*, Weidenfeld & Nicolson.

Bryan, D. (2015). *Cosmos, Chaos and the Kosher Mentality*, Bloomsbury Publishing.

Candela, P. A. (2005). 'Ores in the Earth's crust', *The Crust*, ed. R. L. Rudnick, Elsevier.

Cane, M. A. and P. Molnar (2001). 'Closing of the Indonesian seaway as a precursor to east African aridification around 3–4 million years ago', *Nature* 411: 157–62.

Cann, J. and K. Gillis (2004). 'Hydrothermal insights from the Troodos ophiolite, Cyprus', *Hydrogeology of the Oceanic Lithosphere,* ed. E. E. Davis and H. Elderfield, Cambridge University Press: 274–310.

Cannat, M., A. Briais, C. Deplus, J. Escartin, J. Georgen, J. Lin, S. Mercouriev, C. Meyzen, M. Muller, G. Pouliquen, A. Rabain and P. da Silva (1999). 'Mid-Atlantic Ridge-Azores hotspot interactions: along-axis migration of a hotspot-derived event of enhanced magmatism 10 to 3 Ma ago', *Earth and Planetary Science Letters* 173(3): 257–69.

Carotenuto, F., N. Tsikaridze, L. Rook, D. Lordkipanidze, L. Longo, S. Condemi and P. Raia (2016). 'Venturing out safely: The biogeography of Homo erectus dispersal out of Africa', *Journal of Human Evolution* 95: 1–12.

Castree, N., D. Demeritt, D. Liverman and B. Rhoads (2009). *A Companion to Environmental Geography,* Wiley-Blackwell.

Chamberlin, J. E. (2013). *Island: How Islands Transform the World,* Elliott & Thompson.

Chen, Z. Q. and M. J. Benton (2012). 'The timing and pattern of biotic recovery following the end-Permian mass extinction', *Nature Geoscience* 5(6): 375–83.

Chorowicz, J. (2005). 'The East African rift system', *Journal of African Earth Sciences* 43(1): 379–410.

Clapper, J. R. (2013). *Worldwide Threat Assessment of the US Intelligence Community,* Office of the Director of National Intelligence, Washington, DC.

Clift, P. D., K. V. Hodges, D. Heslop, R. Hannigan, H. Van Long and G. Calves (2008). 'Correlation of Himalayan exhumation rates and Asian monsoon intensity', *Nature Geoscience* 1(12): 875–80.

Constantinou, G. (1982). 'Geological features and ancient exploitation of the cupriferous sulphide orebodies of Cyprus', *Early Metallurgy in Cyprus, 4000–500 BC,* ed. J. D. Muhly, R. Maddin and V. Karageorghis, Pierides Foundation, Nicosia: 13–23.

Corones, M. (2015). 'Mapping world oil transport', http://blogs.reuters.com/data-dive/2015/03/27/mapping-world-oil-transport/.

Cowie, J. (2012). *Climate Change: Biological and Human Aspects*, Cambridge University Press.

Crowley, R. (2016). *Conquerors: How Portugal Forged the First Global Empire*, Faber & Faber.

Crutzen, P. J. and E. F. Stoermer (2000). 'The "Anthropocene"', *International Geosphere–Biosphere Programme (IGBP) Newsletter* 41: 17–18.

Cunningham, C. G., R. E. Zartman, E. H. McKee, R. O. Rye, C. W. Naeser, O. Sanjines, G. E. Ericksen and F. Tavera (1996). 'The age and thermal history of Cerro Rico de Potosi, Bolivia', *Mineralium Deposita* 31(5): 374–85.

Curry, A. (2013). 'Archaeology: The milk revolution', *Nature* 500: 20–2.

Dalvi, S. (2015). *Fundamentals of Oil & Gas Industry for Beginners*, Notion Press.

Dartnell, L. (2015). *The Knowledge: How to Rebuild Our World after an Apocalypse*, Vintage.

De Ryck, I., A. Adriaens and F. Adams (2005). 'An overview of Mesopotamian bronze metallurgy during the 3rd millennium BC', *Journal of Cultural Heritage* 6(3): 261–8.

Diamond, J. (1998). *Guns, Germs and Steel: A Short History of Everybody for the Last 13,000 Years*, Vintage.

Diamond, J. (2011). *Collapse: How Societies Choose to Fail or Survive*, Penguin.

Diamond, J. and P. Bellwood (2003). 'Farmers and their languages: The first expansions', *Science* 300(5619): 597–603.

Douglas, I. and N. Lawson (2000). 'The human dimensions of geo-morphological work in Britain', *Journal of Industrial Ecology* 4(2): 9–33.

Dutch, S. (2002). 'Geology and Election 2000', http://www.uwgb.edu/dutchs/Research/Elec2000/GeolElec2000.HTML/.

Dutch, S. (2006). 'What If? The Ice Ages Had Been A Little Less Icy?', *Geological Society of America meeting abstracts, Philadelphia, PA* 38(7): 73–5.

East, W. G. (1967). *The Geography behind History*, W.W. Norton & Co.

Eriksson, A., L. Betti, A. D. Friend, S. J. Lycett, J. S. Singarayer, N. von Cramon-Taubadel, P. J. Valdes, F. Balloux and A. Manica (2012). 'Late Pleistocene climate change and the global expansion of anatomically modern humans', *Proceedings of the National Academy of Sciences* 109(40): 16089.

Erisman, J. W., M. A. Sutton, J. Galloway, Z. Klimont and W. Winiwarter (2008). 'How a century of ammonia synthesis changed the world', *Nature Geoscience* 1(10): 636–9.

Ermini, L., C. D. Sarkissian, E. Willerslev and L. Orlando (2015). 'Major transitions in human evolution revisited: A tribute to ancient DNA', *Journal of Human Evolution* 79: 4–20.

Fagan, B. (2001). *The Little Ice Age: How Climate Made History 1300–1850*, Basic Books.

Falkowski, P. G. (2015). *Life's Engines: How Microbes Made Earth Habitable*, Princeton University Press.

Fer, I., B. Tietjen, F. Jeltsch and M. H. Trauth (2017). 'Modelling vegetation change during Late Cenozoic uplift of the East African plateaus', *Palaeogeography Palaeoclimatology Palaeoecology* 467: 120–30.

Fernandez-Armesto, F. (2002). *Civilizations: Culture, Ambition, and the Transformation of Nature*, Free Press.

Feurdean, A., S. A. Bhagwat, K. J. Willis, H. J. B. Birks, H. Lischke and T. Hickler (2013). 'Tree Migration-Rates: Narrowing the Gap between Inferred Post-Glacial Rates and Projected Rates', *PLoS One* 8(8): e71797.

Fish, S. (2010). *The Manila-Acapulco Galleons: The Treasure Ships of the Pacific*, AuthorHouse.

Fokkens, H. and A. Harding (2013). *The Oxford Handbook of the European Bronze Age*, Oxford University Press.

Force, E. R. (2008). 'Tectonic environments of ancient civilizations in the eastern hemisphere', *Geoarchaeology* 23(5): 644–53.

Force, E. R. (2015). *Impact of Tectonic Activity on Ancient Civilizations: Recurrent Shakeups, Tenacity, Resilience, and Change*, Lexington Books.

Force, E. R. and B. G. McFadgen (2010). 'Tectonic environments of ancient civilizations: Opportunities for archaeoseismological and anthropological studies', *Ancient Earthquakes*, ed. M. Sintubin, I. S. Stewart, T. M. Niemi and E. Altunel, Geological Society of America.

Force, E. R. and B. G. McFadgen (2012). *Influences of Active Tectonism on Human Development: A Review and Neolithic Example*, Climates, Landscapes, and Civilizations (Geophysical Monograph Series 198), American Geophysical Union.

Fortey, R. (2005). *The Earth: An Intimate History*, Harper Perennial.

Fortey, R. (2010). *The Hidden Landscape: A Journey into the Geological Past*, Bodley Head.

Frankopan, P. (2016). *The Silk Roads: A New History of the World*, Bloomsbury.

Franks, J. W. (1960). 'Interglacial Deposits at Trafalgar Square, London', *New Phytologist* 59(2): 145–52.

Friedman, G. (2017). 'There are 2 choke points that threaten oil trade between the Persian Gulf and East Asia', https://www.businessinsider.com/maps-oil-trade-choke-points-person-gulf-and-east-asia-2017-4?IR=T.

Fromkin, D. (2000). *Way of the World*, Vintage.

Fuller, D. Q., T. Denham, M. Arroyo-Kalin, L. Lucas, C. J. Stevens, L. Qin, R. G. Allaby and M. D. Purugganan (2014). 'Convergent evolution and parallelism in plant domestication revealed by an expanding archaeological record', *Proceedings of the National Academy of Sciences of the United States of America* 111(17): 6147–52.

Garcia-Castellanos, D., F. Estrada, I. Jiménez-Munt, C. Gorini, M. Fernàndez, J. Vergés and R. De Vicente (2009). 'Catastrophic flood of the Mediterranean after the Messinian salinity crisis', *Nature* 462: 778.

Garzanti, E., A. I. Al-Juboury, Y. Zoleikhaei, P. Vermeesch, J. Jotheri, D. B. Akkoca, A. K. Obaid, M. B. Allen, S. Ando, M. Limonta, M. Padoan, A. Resentini, M. Rittner and G. Vezzoli (2016). 'The Euphrates–Tigris–Karun river system: Provenance, recycling and dispersal of quartz-poor foreland-basin sediments in arid climate', *Earth-Science Reviews* 162: 107–28.

Gatti, E. and C. Oppenheimer (2012). *Utilization of Distal Tephra Records for Understanding Climatic and Environmental Consequences of the Youngest Toba Tuff*. Climates, Landscapes, and Civilizations (Geophysical Monograph Series 198), American Geophysical Union.

Gehler, A., P. D. Gingerich and A. Pack (2016). 'Temperature and atmospheric CO_2 concentration estimates through the PETM using triple oxygen isotope analysis of mammalian bioapatite', *Proceedings of the National Academy of Sciences of the United States of America* 113(28): 7739–44.

Gibbard, P. (2007). 'Europe cut adrift', *Nature* 448: 259.

Gibbons, A. (1998). 'Ancient island tools suggest Homo erectus was a seafarer', *Science* 279(5357): 1635–7.

Gingerich, P. D. (2006). 'Environment and evolution through the Paleocene-Eocene thermal maximum', *Trends in Ecology & Evolution* 21(5): 246–53.

Giosan, L., P. D. Clift, M. G. Macklin, D. Q. Fuller, S. Constantinescu, J. A. Durcan, T. Stevens, G. A. T. Duller, A. R. Tabrez, K. Gangal, R. Adhikari, A. Alizai, F. Filip, S. VanLaningham and J. P. M. Syvitski (2012). 'Fluvial landscapes of the Harappan civilization', *Proceedings of the National Academy of Sciences of the United States of America* 109(26): E1688–94.

Goody, J. (2012). *Metals, Culture and Capitalism: An Essay on the Origins of the Modern World*, Cambridge University Press.

Graedel, T. E., G. Gunn and L. T. Espinoza (2014). 'Metal resources, use and criticality', *Critical Metals Handbook*, ed. G. Gunn, AGU/Wiley: 1–19.

Greene, K. (2000). 'Technological innovation and economic progress in the ancient world: M. I. Finley re-considered', *Economic History Review* 53(1): 29–59.

Gregory, K. J. (2010). *The Earth's Land Surface: Landforms and Processes in Geomorphology*, SAGE Publications.

Grosman, L., N. D. Munro and A. Belfer-Cohen (2008). 'A 12,000-year-old Shaman burial from the southern Levant (Israel)', *Proceedings of the National Academy of Sciences of the United States of America* 105(46): 17665–9.

Guimaraes, P. R., M. Galetti and P. Jordano (2008). 'Seed Dispersal Anachronisms: Rethinking the Fruits Extinct Megafauna Ate', *PLoS One* 3(3).

Guinotte, J. M. and V. J. Fabry (2008). 'Ocean Acidification and Its Potential Effects on Marine Ecosystems', *Annals of the New York Academy of Sciences* 1134(1): 320–42.

Gunn, G. (2014). 'Platinum-group metals', *Critical Metals Handbook*, ed. G. Gunn, AGU/Wiley: 284–311.

Gupta, S., J. S. Collier, D. Garcia-Moreno, F. Oggioni, A. Trentesaux, K. Vanneste, M. De Batist, T. Camelbeeck, G. Potter, B. Van Vliet-Lanoe and J. C. R. Arthur (2017). 'Two-stage opening of the Dover Strait and the origin of island Britain', *Nature Communications* 8: 1–12.

Gupta, S., J. S. Collier, A. Palmer-Felgate and G. Potter (2007). 'Catastrophic flooding origin of shelf valley systems in the English Channel', *Nature* 448(7151): 342–6.

Haase, D., J. Fink, G. Haase, R. Ruske, M. Pecsi, H. Richter, M. Altermann and K. D. Jager (2007). 'Loess in Europe – its spatial distribution based on a European Loess Map, scale 1:2,500,000', *Quaternary Science Reviews* 26(9–10): 1301–12.

Hagelüken, C. (2014). 'Recycling of (critical) metals', *Critical Metals Handbook*, ed. G. Gunn, AGU/Wiley: 41–69.

Hamilton, T. L., D. A. Bryant and J. L. Macalady (2016). 'The role of biology in planetary evolution: cyanobacterial primary production in low-oxygen Proterozoic oceans', *Environmental Microbiology* 18(2): 325–40.

Hanson, T. (2016). *The Triumph of Seeds: How Grains, Nuts, Kernels, Pulses, and Pips Conquered the Plant Kingdom and Shaped Human History*, Basic Books.

Haywood, J. (2012). *Chronicles of the Ancient World*, Quercus.

Headrick, D. R. (2010). *Power over Peoples: Technology, Environments, and Western Imperialism, 1400 to the Present*, Princeton University Press.

Helly, J. J. and L. A. Levin (2004). 'Global distribution of naturally occurring marine hypoxia on continental margins', *Deep Sea Research Part I: Oceanographic Research Papers* 51(9): 1159–68.

Henrich, J. (2004). 'Demography and Cultural Evolution: How Adaptive Cultural Processes can Produce Maladaptive Losses: The Tasmanian Case', *American Antiquity* 69(2): 197–214.

Hillstrom, K. and L. C. Hillstrom (2005). *Industrial Revolution in America: Iron and Steel*, ABC-CLIO.

Hodell, D. A., J. H. Curtis and M. Brenner (1995). 'Possible role of climate in the collapse of Classic Maya civilization', *Nature* 375: 391–4.

Hoffecker, J. F. (2005). 'Innovation and technological knowledge in the upper paleolithic of northern Eurasia', *Evolutionary Anthropology* 14(5): 186–98.

Hoffman, P. T. (2017). *Why Did Europe Conquer the World?*, Princeton University Press.

Holen, S. R., T. A. Deméré, D. C. Fisher, R. Fullagar, J. B. Paces, G. T. Jefferson, J. M. Beeton, R. A. Cerutti, A. N. Rountrey, L. Vescera and K. A. Holen (2017). 'A 130,000-year-old archaeological site in southern California, USA', *Nature* 544: 479–83.

Hu, Y. W., H. Shang, H. W. Tong, O. Nehlich, W. Liu, C. H. Zhao, J. C. Yu, C. S. Wang, E. Trinkaus and M. P. Richards (2009). 'Stable isotope dietary analysis of the Tianyuan 1 early modern human', *Proceedings of the National Academy of Sciences of the United States of America* 106(27): 10971–4.

Huang, J. and M. B. McElroy (2014). 'Contributions of the Hadley and Ferrel Circulations to the Energetics of the Atmosphere over the Past 32 Years', *Journal of Climate* 27(7): 2656–66.

Hughey, J. R., P. Paschou, P. Drineas, D. Mastropaolo, D. M. Lotakis, P. A. Navas, M. Michalodimitrakis, J. A. Stamatoyannopoulos and G. Stamatoyannopoulos (2013).'A European population in Minoan Bronze Age Crete', *Nature Communications* 4.

Humphreys, D. (2014). 'The mining industry and the supply of critical minerals', *Critical Metals Handbook*, ed. G. Gunn, AGU/Wiley: 20–40.

Jackson, J. (2006). 'Fatal attraction: living with earthquakes, the growth of villages into megacities, and earthquake vulnerability in the modern world', *Philosophical Transactions of the Royal Society A – Mathematical Physical and Engineering Sciences* 364(1845): 1911–25.

Jacobs, J. (2018). 'Europe's half a million landfill sites potentially worth a fortune', *Financial Times*, https://www.ft.com/content/obf645dc-d8f1-11e7-9504-59efdb70e12f.

Janzen, D. H. and P. S. Martin (1982). 'Neotropical Anachronisms – the Fruits the Gomphotheres Ate', *Science* 215(4528): 19–27.

Ji, R., P. Cui, F. Ding, J. Geng, H. Gao, H. Zhang, J. Yu, S. Hu and H. Meng (2009). 'Monophyletic origin of domestic bactrian camel (Camelus bactrianus) and its evolutionary relationship with the extant wild camel (Camelus bactrianus ferus)', *Animal Genetics* 40(4): 377–82.

Jiao, C., G. Yu, N. He, A. Ma, J. Ge and Z. Hu (2017). 'Spatial pattern of grassland aboveground biomass and its environmental controls in the Eurasian steppe', *Journal of Geographical Sciences* 27(1): 3–22.

Jones, M., R. Jones and M. Woods (2004). *An Introduction to Political Geography: Space, Place and Politics*, Routledge.

Jung, G., M. Prange and M. Schulz (2016). 'Influence of topography on tropical African vegetation coverage', *Climate Dynamics* 46(7): 2535–49.

Kaplan, R. D. (2017). *The Revenge Of Geography,* Random House.

Karol, K. G., R. M. McCourt, M. T. Cimino and C. F. Delwiche (2001). 'The closest living relatives of land plants', *Science* 294(5550): 2351–3.

Kasen, D., B. Metzger, J. Barnes, E. Quataert and E. Ramirez-Ruiz (2017). 'Origin of the heavy elements in binary neutron-star mergers from a gravitational-wave event', *Nature* 551(7678): 80–4.

Kassianidou, V. (2013). 'Mining landscapes of prehistoric Cyprus', *Metalla* 20(2): 5–57.

Keegan, J. (1993). *A History Of Warfare*, Pimlico.

Kenrick, P. and P. R. Crane (1997). 'The origin and early evolution of plants on land', *Nature* 389(6646): 33–9.

Kilian, B., W. Martin and F. Salamini (2010). 'Genetic Diversity, Evolution and Domestication of Wheat and Barley in the Fertile Crescent', *Evolution in Action: Case Studies in Adaptive Radiation, Speciation and the Origin of Biodiversity*, ed. M. Glaubrecht, Springer: 137–66.

Kimber, C. T. (2000). 'Origin of domesticated Sorghum and its early diffusion to India and China', *Sorghum: Origin, History, Technology, and Production*, ed. C. W. Smith, Wiley.

King, G. and G. Bailey (2006). 'Tectonics and human evolution', Antiquity 80(308): 265–86.

King, G., G. Bailey and D. Sturdy (1994). 'Active tectonics and human survival strategies', *Journal of Geophysical Research: Solid Earth* 99(B10): 20063–78.

Kinnaird, J. A. (2005). 'The Bushveld Large Igneous Province', http://www.largeigneousprovinces.org/sites/default/files/BushveldLIP.pdf.

Klein, C. (2005). 'Some Precambrian banded iron-formations (BIFs) from around the world: Their age, geologic setting, mineralogy, metamorphism, geochemistry, and origin', *American Mineralogist* 90(10): 1473–99.

Kleine, T. (2011). 'Earth's patchy late veneer', *Nature* 477(7363): 168–9.

Kneller, B. C. and M. Aftalion (1987). 'The isotopic and structural age of the Aberdeen Granite', *Journal of the Geological Society* 144(5): 717–21.

Koestler-Grack, R. A. (2010). *Mount Rushmore*, ABDO Publishing Co.

Kolarik, Z. and E. V. Renard (2005). 'Potential Applications of Fission Platinoids in Industry', *Platinum Metals Review* 49(2): 79–90.

Kourmpetli, S. and S. Drea (2014). 'The fruit, the whole fruit, and everything about the fruit', *Journal of Experimental Botany* 65(16): 4491–503.

Krause, J., L. Orlando, D. Serre, B. Viola, K. Prüfer, M. P. Richards, J.-J. Hublin, C. Hänni, A. P. Derevianko and S. Pääbo (2007). 'Neanderthals in central Asia and Siberia', *Nature* 449: 902–904.

Krijgsman, W., F. J. Hilgen, I. Raffi, F. J. Sierro and D. S. Wilson (1999). 'Chronology, causes and progression of the Messinian salinity crisis', *Nature* 400: 652–5.

Krivolutskaya, N., B. Gongalsky, A. Dolgal, N. Svirskaya and T. Vekshina (2016). 'Siberian Traps in the Norilsk Area: A Corrected Scheme of Magmatism Evolution', *IOP Conference Series: Earth and Environmental Science* 44: 042008.

Kroeker, K. J., R. L. Kordas, R. N. Crim and G. G. Singh (2010). 'Meta-analysis reveals negative yet variable effects of ocean acidification on marine organisms', *Ecology Letters* 13(11): 1419–34.

Kukula, M. (2016). *The Intimate Universe: How the stars are closer than you think*, Quercus.

Laitin, D. D., J. Moortgat and A. L. Robinson (2012). 'Geographic axes and the persistence of cultural diversity', *Proceedings of the National Academy of Sciences of the United States of America* 109(26): 10263–8.

Larson, G., D. R. Piperno, R. G. Allaby, M. D. Purugganan, L. Andersson, M. Arroyo-Kalin, L. Barton, C. C. Vigueira, T. Denham, K. Dobney, A. N. Doust, P. Gepts, M. T. P. Gilbert, K. J. Gremillion, L. Lucas, L. Lukens, F. B. Marshall, K. M. Olsen, J. C. Pires, P. J. Richerson, R. R. de Casas, O. I. Sanjur, M. G. Thomas and D. Q. Fuller (2014). 'Current perspectives and the future of domestication studies', *Proceedings of the National Academy of Sciences of the United States of America* 111(17): 6139–46.

Larson, R. L. (1991). 'Geological Consequences of Superplumes', *Geology* 19(10): 963–6.

Leidwanger, J., C. Knappett, P. Arnaud, P. Arthur, E. Blake, C. Broodbank, T. Brughmans, T. Evans, S. Graham, E. S. Greene, B. Kowalzig, B. Mills, R. Rivers, T. F. Tartaron and R. V. d. Noort (2014). 'A manifesto for the study of ancient Mediterranean maritime networks', *Antiquity* 88(342).

Lenton, T. and A. Watson (2013). *Revolutions that Made the Earth*, Oxford University Press.

Leveau, P. (1996). 'The Barbegal water mill in its environment: archaeology and the economic and social history of antiquity', *Journal of Roman Archaeology* 9: 137–53.

Lewis, L. (2008). 'There's gold in Japan's landfills', *Sunday Times*. https://www.thetimes.co.uk/article/theres-gold-in-japans-landfills-gfvolwdzh6n.

Lewis, S. L. and M. A. Maslin (2015). 'Defining the Anthropocene', *Nature* 519: 171.

Liddy, H. M., S. J. Feakins and J. E. Tierney (2016). 'Cooling and drying in northeast Africa across the Pliocene', *Earth and Planetary Science Letters* 449: 430–8.

Lieberman, D. (2014). *The Story of the Human Body: Evolution, Health and Disease,* Penguin.

Lim, L. (2004). 'China's drive to transform Tibet', http://news.bbc.co.uk/1/hi/world/asia-pacific/3625588.stm.

Londo, J. P., Y. C. Chiang, K. H. Hung, T. Y. Chiang and B. A. Schaal (2006). 'Phylogeography of Asian wild rice, Oryza rufipogon, reveals multiple independent domestications of cultivated rice, Oryza sativa', *Proceedings of the National Academy of Sciences of the United States of America* 103(25): 9578–83.

López, S., L. van Dorp and G. Hellenthal (2015). 'Human Dispersal Out of Africa: A Lasting Debate', *Evolutionary Bioinformatics Online* 11(Suppl 2): 57–68.

Lutgens, F. K. and E. J. Tarbuck (2000). *The Atmosphere: An Introduction to Meteorology*, 8th edition, Prentice Hall.

Lyons, T. W., C. T. Reinhard and N. J. Planavsky (2014). 'The rise of oxygen in Earth's early ocean and atmosphere', *Nature* 506: 307–15.

Macalister, T. (2015). 'Kellingley colliery closure:"shabby end" for a once mighty industry', *Guardian*, https://www.theguardian.com/environment/2015/dec/18/kellingley-colliery-shabby-end-for-an-industry.

Mann, P., L. Gahagan and M. B. Gordon (2003). 'Tectonic setting of the world's giant oil and gas fields', *Giant Oil and Gas Fields of the Decade 1990–1999*, ed. M. T. Halbouty, AAPG Memoir 78: 15–105.

Marr, A. (2013). *A History of the World*, Pan.

Marshall, T. (2016). *Prisoners of Geography: Ten Maps That Tell You Everything You Need to Know About Global Politics*, Elliott & Thompson.

Maslin, M. (2013). 'How a changing landscape and climate shaped early humans', https://theconversation.com/how-a-changing-landscape-and-climate-shaped-early-humans-19862.

Maslin, M. A., C. M. Brierley, A. M. Milner, S. Shultz, M. H. Trauth and K. E. Wilson (2014). 'East African climate pulses and early human evolution', *Quaternary Science Reviews* 101: 1–17.

Maslin, M. A. and B. Christensen (2007). 'Tectonics, orbital forcing, global climate change, and human evolution in Africa: introduction to the African paleoclimate special volume', *Journal of Human Evolution* 53(5): 443–64.

Mayewski, P. A., E. E. Rohling, J. C. Stager, W. Karlen, K. A. Maasch, L. D. Meeker, E. A. Meyerson, F. Gasse, S. van Kreveld, K. Holmgren, J. Lee-Thorp, G. Rosqvist, F. Rack, M. Staubwasser, R. R. Schneider and E. J. Steig (2004). 'Holocene climate variability', *Quaternary Research* 62(3): 243–55.

McBrearty, S. and A. S. Brooks (2000). 'The revolution that wasn't: a new interpretation of the origin of modern human behavior', *Journal of Human Evolution* 39(5): 453–563.

McCormick, M., U. Buntgen, M. A. Cane, E. R. Cook, K. Harper, P. Huybers, T. Litt, S. W. Manning, P. A. Mayewski, A. F. M. More, K. Nicolussi and W. Tegel (2012). 'Climate Change during and after the Roman Empire: Reconstructing the Past from Scientific and Historical Evidence', *Journal of Interdisciplinary History* 43(2): 169–220.

McDermott, R. (1998). *Risk-Taking in International Politics: Prospect Theory in American Foreign Policy*, University of Michigan Press.

McDougall, E. A. (1983). 'The Sahara Reconsidered: Pastoralism, Politics and Salt from the Ninth through the Twelfth Centuries', *African Economic History* 12: 263–86.

McDougall, E. A. (1990). 'Salts of the Western Sahara: Myths, Mysteries, and Historical Significance', *International Journal of African Historical Studies* 23(2): 231–57.

McInerney, F. A. and S. L. Wing (2011). 'The Paleocene-Eocene Thermal Maximum: A Perturbation of Carbon Cycle, Climate, and Biosphere with Implications for the Future', *Annual Review of Earth and Planetary Sciences*, 39: 489–516.

McKie, R. (2013). 'Why did the Neanderthals die out?' *Observer*, https://www.theguardian.com/science/2013/jun/02/why-did-neanderthals-die-out.

McNeill, J. R. (2001). 'The World According to Jared Diamond', *The History Teacher* 34(2): 165–74.

McNeill, J. R. (2012a). 'Global Environmental History: The First 150,000 Years', *A Companion to Global Environmental History*, ed. J. R. McNeill and E. S. Mauldin, Blackwell Publishing: 3–17.

McNeill, J. R. (2012b). 'Biological Exchange in Global Environmental History', *A Companion to Global Environmental History*, ed. J. R. McNeill, Blackwell Publishing: 433–51.

McNeill, J. R. and W. H. McNeill (2004). *The Human Web: A Bird's-Eye View of World History*, W. W. Norton & Co.

McNeill, W. H. (1963). *The Rise of the West: A History of the Human Community*, University of Chicago Press.

Mendez, A. (2011). 'Distribution of landmasses of the Paleo-Earth', http://phl.upr.edu/library/notes/distributionoflandmassesofthepaleo-earth.

Metspalu, M., T. Kivisild, E. Metspalu, J. Parik, G. Hudjashov, K. Kaldma, P. Serk, M. Karmin, D. M. Behar, M. T. P. Gilbert, P. Endicott, S. Mastana, S. S. Papiha, K. Skorecki, A. Torroni and R. Villems (2004). 'Most of the extant mtDNA boundaries in South and Southwest Asia were likely shaped during the initial settlement of Eurasia by anatomically modern humans', *BMC Genetics* 5: 26.

Millward, J. A. (2013). *The Silk Road: A Very Short Introduction*, Oxford University Press.

Mokyr, J. (1992). *The Lever of Riches: Technological Creativity and Economic Progress*, Oxford University Press.

Monbiot, G. (2014). *Feral: Rewilding the Land, Sea and Human Life*, Penguin.

Moorjani, P., C. E. G. Amorim, P. F. Arndt and M. Przeworski (2016). 'Variation in the molecular clock of primates', *Proceedings of the National Academy of Sciences* 113(38): 10607–12.

Morris, I. (2011). *Why the West Rules – for Now: The Patterns of History and What They Reveal about the Future*, Profile Books.

Morton, O. (2016). *The Planet Remade: How Geoengineering Could Change the World*, Granta.

Moylan, J. (2017). 'First coal-free day in Britain since 1880s', https://www.bbc.co.uk/news/uk-39675418.

Murton, J. B., M. D. Bateman, S. R. Dallimore, J. T. Teller and Z. R. Yang (2010). 'Identification of Younger Dryas outburst flood path from Lake Agassiz to the Arctic Ocean', *Nature* 464(7289): 740–3.

Myers, J. S. (1997). 'Geology of granite', *Journal of the Royal Society of Western Australia* 80: 87–100.

Needham, J. (1965). *Science and Civilisation in China. Volume 4: Physics and Physical Technology. Part II: Mechanical Engineering*, Cambridge University Press.

Neimark, J. (2012). 'How We Won the Hominid Wars, and All the Others Died Out', *Discover*, http://discovermagazine.com/2011/evolution/23-how-we-won-the-hominid-wars.

Nelsen, M. P., W. A. DiMichele, S. E. Peters and C. K. Boyce (2016). 'Delayed fungal evolution did not cause the Paleozoic peak in coal production', *Proceedings of the National Academy of Sciences* 113(9): 2442–7.

Nicoll, K. (2013). *Geoarchaeological Perspectives on Holocene Climate Change as a Civilizing Factor in the Egyptian Sahara*, Climates, Landscapes, and Civilizations (Geophysical Monograph Series 198), American Geophysical Union.

Nield, T. (2014). *Underlands: A Journey through Britain's Lost Landscape*, Granta.

Novacek, M. (2008). *Terra: Our 100 Million Year Old Ecosystem and the Threats That Now Put It at Risk*, Farrar, Straus and Giroux.

O'Dea, A., H. A. Lessios, A. G. Coates, R. I. Eytan, S. A. Restrepo-Moreno, A. L. Cione, L. S. Collins, A. de Queiroz, D. W. Farris, R. D. Norris, R. F. Stallard, M. O. Woodburne, O. Aguilera, M.-P. Aubry, W. A. Berggren, A. F. Budd, M. A. Cozzuol, S. E. Coppard, H. Duque-Caro, S. Finnegan, G. M. Gasparini, E. L. Grossman, K. G. Johnson, L. D. Keigwin, N. Knowlton, E. G. Leigh, J. S. Leonard-Pingel, P. B. Marko, N. D. Pyenson, P. G. Rachello-Dolmen, E. Soibelzon, L. Soibelzon, J. A. Todd, G. J. Vermeij and J. B. C. Jackson (2016). 'Formation of the Isthmus of Panama', *Science Advances* 2(8): e1600883.

Oleson, J. P. (2009). *The Oxford Handbook of Engineering and Technology in the Classical World*, Oxford University Press.

Oppenheimer, C. (2011). *Eruptions that Shook the World*, Cambridge University Press.

Orlando, L. (2016). 'Back to the roots and routes of dromedary domestication', *Proceedings of the National Academy of Sciences of the United States of America* 113(24): 6588–90.

Osborne, R. (2013). *Iron, Steam & Money: The Making of the Industrial Revolution*, Bodley Head.

Paine, L. (2013). *The Sea and Civilization: A Maritime History of the World*, Knopf.

Parker, G. (1976). 'The 'Military Revolution', 1560–1660 – a Myth?', *Journal of Modern History* 48(2): 196–214.

Patterson, N., D. J. Richter, S. Gnerre, E. S. Lander and D. Reich (2006). 'Genetic evidence for complex speciation of humans and chimpanzees', *Nature* 441: 1103–8.

Petit, J. R., J. Jouzel, D. Raynaud, N. I. Barkov, J. M. Barnola, I. Basile, M. Bender, J. Chappellaz, M. Davis, G. Delaygue, M. Delmotte, V. M. Kotlyakov, M. Legrand, V. Y. Lipenkov, C. Lorius, L. Pepin, C. Ritz, E. Saltzman and M. Stievenard (1999). 'Climate and atmospheric history of the past 420,000 years from the Vostok ice core, Antarctica', *Nature* 399(6735): 429–36.

Phillips, W. R. (1988). 'Ancient Civilizations and Geology of the Eastern Mediterranean', *Excavations at Seila, Egypt*, ed. C. W. Griggs, Brigham Young University: 1–18.

Piantadosi, C. A. (2003). *The Biology of Human Survival: Life and Death in Extreme Environments*, Oxford University Press.

Pim, J., C. Peirce, A. B. Watts, I. Grevemeyer and A. Krabbenhoeft (2008). 'Crustal structure and origin of the Cape Verde Rise', *Earth and Planetary Science Letters* 272(1–2): 422–8.

Pollard, J. (2017). 'The Uffington White Horse geoglyph as sun-horse', *Antiquity* 91(356): 406–20.

Potts, R. (2013). 'Hominin evolution in settings of strong environmental variability', *Quaternary Science Reviews* 73: 1–13.

Potts, R. and J. T. Faith (2015). 'Alternating high and low climate variability: The context of natural selection and speciation in Plio-Pleistocene hominin evolution', *Journal of Human Evolution* 87: 5–20.

Price, T. D. (2009). 'Ancient farming in eastern North America', *Proceedings of the National Academy of Sciences of the United States of America* 106(16): 6427–8.

Pye, K. (1995). 'The nature, origin and accumulation of loess', *Quaternary Science Reviews* 14(7–8): 653–67.

Pye, M. (2015). *The Edge of the World: How the North Sea Made Us Who We Are*, Penguin.

Qiu, J. (2014). 'Double threat for Tibet', *Nature* 512: 240–1.

Rackham, O. (2009). 'Ancient Forestry Practices', *The Role of Food,*

Agriculture, Forestry and Fisheries in Human Nutrition, Volume II, ed. V. R. Squires, EOLSS Publishers: 29–47.

Raj, N. G. (2013). 'The Tibetan plateau and the Indian monsoon', https://www.thehindu.com/sci-tech/science/the-tibetan-plateau-and-the-indian-monsoon/article4651084.ece.

Rajagopalan, B. and P. Molnar (2013). 'Signatures of Tibetan Plateau heating on Indian summer monsoon rainfall variability', *Journal of Geophysical Research-Atmospheres* 118(3): 1170–8.

Ramachandran, S. and N. A. Rosenberg (2011). 'A Test of the Influence of Continental Axes of Orientation on Patterns of Human Gene Flow', *American Journal of Physical Anthropology* 146(4): 515–29.

Ramalho, R., G. Helffrich, D. N. Schmidt and D. Vance (2010). 'Tracers of uplift and subsidence in the Cape Verde archipelago', *Journal of the Geological Society* 167(3): 519–38.

Rasmussen, S. C. (2012). *How Glass Changed the World: The History and Chemistry of Glass from Antiquity to the 13th Century*, Springer.

Raudzens, G. (2003). *Technology, Disease, and Colonial Conquests, Sixteenth to Eighteenth Centuries: Essays Reappraising the Guns and Germs Theories*, Brill.

Reader, J. (2005). *Cities*, Vintage.

Reilinger, R. and S. McClusky (2011). 'Nubia-Arabia-Eurasia plate motions and the dynamics of Mediterranean and Middle East tectonics', *Geophysical Journal International* 186(3): 971–9.

Richerson, P. J., R. Boyd and R. L. Bettinger (2001). 'Was Agriculture Impossible during the Pleistocene but Mandatory during the Holocene? A Climate Change Hypothesis', *American Antiquity* 66(3): 387–411.

Rick, T. C. and J. M. Erlandson (2008). *Human Impacts on Ancient Marine Ecosystems: A Global Perspective*, University of California Press.

Ridley, J. (2013). *Ore Deposit Geology*, Cambridge University Press.

Roberts, B. W., C. P. Thornton and V. C. Pigott (2009). 'Development of metallurgy in Eurasia', *Antiquity* 83(322): 1012–22.

Roberts, M. (1967). *Essays in Swedish History*, London: Weidenfeld & Nicolson.

Rodger, N. A. M. (2012). 'Atlantic Seafaring', *The Oxford Handbook of the Atlantic World: 1450–1850*, ed. C. Nicholas and M. Philip, Oxford University Press.

Roebroeks, W., M. J. Sier, T. K. Nielsen, D. De Loecker, J. M. Pares, C. E. S. Arps and H. J. Mucher (2012). 'Use of red ochre by early Neandertals', *Proceedings of the National Academy of Sciences of the United States of America* 109(6): 1889–94.

Rogers, C. (2018). *The Military Revolution Debate: Readings on the Military Transformation of Early Modern Europe*, Routledge.

Rohrig, B. (2015). 'Smartphones: Smart Chemistry', https://www.acs.org/content/acs/en/education/resources/highschool/chemmatters/past-issues/archive-2014-2015/smartphones.html.

Rose, J. I., V. I. Usik, A. E. Marks, Y. H. Hilbert, C. S. Galletti, A. Parton, J. M. Geiling, V. Cerny, M. W. Morley and R. G. Roberts (2011). 'The Nubian Complex of Dhofar, Oman: An African Middle Stone Age Industry in Southern Arabia', *PloS One* 6(11).

Rosenbaum, G., G. S. Lister and C. Duboz (2002). 'Reconstruction of the tectonic evolution of the western Mediterranean since the Oligocene', *Journal of the Virtual Explorer* 8: 107–30.

Rothery, D. (2010). *Geology: The Key Ideas*, Teach Yourself.

Ruddiman, W. F. (2016). *Plows, Plagues, and Petroleum: How Humans Took Control of Climate*, Princeton University Press.

Ruddiman, W. F., E. C. Ellis, J. O. Kaplan and D. Q. Fuller (2015). 'Defining the epoch we live in." *Science* 348(6230): 38–9.

Sadykov, V. A., L. A. Isupova, I. A. Zolotarskii, L. N. Bobrova, A. S. Noskov, V. N. Parmon, E. A. Brushtein, T. V. Telyatnikova, V. I. Chernyshev and V. V. Lunin (2000). 'Oxide catalysts for ammonia oxidation in nitric acid production: properties and perspectives', *Applied Catalysis A: General* 204(1): 59–87.

Sagan, C. (1973). *The Cosmic Connection: An Extraterrestrial Perspective*, Doubleday.

Sage, R. F. (1995). 'Was low atmospheric CO_2 during the Pleistocene a limiting factor for the origin of agriculture?', *Global Change Biology* 1(2): 93–106.

Sahrhage, D. and J. Lundbeck (2012). *A History of Fishing*, Springer.

Sakellariou, D. and N. Galanidou (2016). 'Pleistocene submerged landscapes and Palaeolithic archaeology in the tectonically active Aegean region', *Geology and Archaeology: Suberged Landscapes of the Continental Shelf. Geological Society Special Publication No. 411*, ed. J. Harff, G. Bailey and F. Lüth, Geological Society.

Sauberlich, H. E. (1997). 'A History of Scurvy and Vitamin C', *Vitamin C in Health and Disease*, ed. L. Packer, Taylor & Francis.

Schmidt, M. (2017). 'Human Geography of Post-Socialist Mountain Regions: An Introduction', *Journal of Alpine Research* 105(1).

Schobert, H. H. (2014). *Energy and Society: An Introduction*, 2nd edition, CRC Press.

Schrijver, K. and I. Schrijver (2015). *Living with the Stars: How the Human Body is Connected to the Life Cycles of the Earth, the Planets, and the Stars*, Oxford University Press.

Schuberth, C. J. (1968). *The Geology of New York City and Environs*, Natural History Press for the American Museum of Natural History.

Schwarz-Schampera, U. (2014). 'Indium', *Critical Metals Handbook*, ed. G. Gunn, AGU/Wiley: 204–29.

Scott, A. C. and I. J. Glasspool (2006). 'The diversification of Paleozoic fire systems and fluctuations in atmospheric oxygen concentration', *Proceedings of the National Academy of Sciences* 103(29): 10861–5.

Shakun, J. D., P. U. Clark, F. He, S. A. Marcott, A. C. Mix, Z. Y. Liu, B. Otto-Bliesner, A. Schmittner and E. Bard (2012). 'Global warming preceded by increasing carbon dioxide concentrations during the last deglaciation', *Nature* 484(7392): 49–54.

Sheridan, A. (2002). 'ANTIQUITY and the Old World', *Antiquity* 76(294): 1085–8.

Shubin, N. (2014). *The Universe Within*, Penguin.

Shuckburgh, E. and P. Austin (2008). *Survival: The Survival of the Human Race*, Cambridge University Press.

Siddall, R. (2015). 'An Urban Geologist's Guide to the Fossils of the Portland Stone, Urban Geology in London No. 30', http://www.ucl.ac.uk/~ucfbrxs/Homepage/walks/PortlandFossils.pdf.

Sinha, U. K. (2010). 'Tibet's watershed challenge', *Washington Post*, 14 June, http://www.washingtonpost.com/wp-dyn/content/article/2010/06/ 13/AR2010061303331.html.

Smith, A. H. V. (1997). 'Provenance of Coals from Roman Sites in England and Wales', *Britannia* 28: 297–324.

Smith, B. D. and R. A. Yarnell (2009). 'Initial formation of an indigenous crop complex in eastern North America at 3800 B.P.', *Proceedings of the National Academy of Sciences* 106(16): 6561–6.

Socratous, M. A., V. Kassianidou and G. D. Pasquale (2011). 'Ancient slag heaps in Cyprus: The contribution of charcoal analysis to the study of the ancient copper industry', *Archaeometallurgy in Europe*

III, ed. A. Hauptmann and D. Modarressi-Tehrani. Deutsches Bergbau-Museum Bochum.

Sorkhabi, R. (2016). 'Rich Petroleum Source Rocks', *GEO ExPro* 6(6).

Speight, J. G. (2015). *Asphalt Materials Science and Technology,* Butterworth-Heinemann.

Stager, C. (2012). *Our Future Earth,* Gerald Duckworth & Co.

Stager, C. (2014). *Your Atomic Self: The Invisible Elements That Connect You to Everything Else in the Universe,* Thomas Dunne.

Stahl, P. W. (2008). 'Animal Domestication in South America', *The Handbook of South American Archaeology,* ed. H. Silverman and W. H. Isbell, Springer: 121–30.

Stampfli, G. M., C. Hochard, C. Vérard, C. Wilhem and J. vonRaumer (2013). 'The formation of Pangea', *Tectonophysics* 593: 1–19.

Stavridis, J. (2018). *Sea Power: The History and Geopolitics of the World's Oceans,* Penguin.

Sterelny, K. (2011). 'From hominins to humans: how sapiens became behaviourally modern', *Philosophical Transactions of the Royal Society B – Biological Sciences* 366(1566): 809–22.

Stern, R. J. (2010). 'United States cost of military force projection in the Persian Gulf, 1976–2007', *Energy Policy* 38(6): 2816–25.

Stewart, K. M. (1994). 'Early hominid utilisation of fish resources and implications for seasonality and behaviour', *Journal of Human Evolution* 27: 229–45.

Stewart, W. M., D. W. Dibb, A. E. Johnston and T. J. Smyth (2005). 'The contribution of commercial fertilizer nutrients to food production', *Agronomy Journal* 97(1): 1–6.

Stoneley, R. (1990). 'The Middle East Basin: a summary overview', *Classic Petroleum Provinces, Geological Society Special Publication No. 50,* ed. J. Brooks, Geological Society of London: 293–8.

Stow, D. (2010). *Vanished Ocean: How Tethys Reshaped the World,* Oxford University Press.

Summerhayes, C. P. (2015). *Earth's Climate Evolution,* Wiley Blackwell.

Sweeney, E. J. (2007). *The Pyramid Age,* Algora.

Tarasov, L. and W. R. Peltier (2005). 'Arctic freshwater forcing of the Younger Dryas cold reversal', *Nature* 435: 662.

Teller, J. T., D. W. Leverington and J. D. Mann (2002). 'Freshwater outbursts to the oceans from glacial Lake Agassiz and their role in climate change during the last deglaciation', *Quaternary Science Reviews* 21(8): 879–87.

Terazono, E. (2016). 'Russia set to be biggest wheat exporter for first time', *Financial Times*, https://www.ft.com/content/af66f51e-6515-11e6-8310-ecfobddad227.

Thomas, L. (2013). *Coal Geology*, 2nd edition, Wiley.

Thompson, P. (2010). *Seeds, Sex and Civilization: How the Hidden Life of Plants Has Shaped Our World*, Thames & Hudson.

Tomczak, M. and J. S. Godfrey (1994). *Regional Oceanography: An Introduction*, Pergamon.

Törnqvist, T. E. and M. P. Hijma (2012). 'Links between early Holocene ice-sheet decay, sea-level rise and abrupt climate change', *Nature Geoscience* 5: 601–606.

Trauth, M. H., M. A. Maslin, A. L. Deino, A. Junginger, M. Lesoloyia, E. O. Odada, D. O. Olago, L. A. Olaka, M. R. Strecker and R. Tiedemann (2010). 'Human evolution in a variable environment: the amplifier lakes of Eastern Africa', *Quaternary Science Reviews* 29(23–24): 2981–8.

Trauth, M. H., M. A. Maslin, A. L. Deino, M. R. Strecker, A. G. N. Bergner and M. Duhnforth (2007). 'High- and low-latitude forcing of Plio-Pleistocene East African climate and human evolution', *Journal of Human Evolution* 53(5): 475–86.

Tucci, S. and J. M. Akey (2016). 'A map of human wanderlust', *Nature* 538: 179–80.

Twining, D. (2009). 'Could China and India go to war over Tibet?', https://foreignpolicy.com/2009/03/10/could-china-and-india-go-to-war-over-tibet/.

Ulmishek, G. F. and H. D. Klemme (1990). 'Depositional controls, distribution, and effectiveness of world's petroleum source rocks', *US Geological Survey Bulletin* 1931, US Geological Survey.

Vasiljevic, D. A., S. B. Markovic, T. A. Hose, Z. L. Ding, Z. T. Guo, X. M. Liu, I. Smalley, T. Lukic and M. D. Vujicic (2014). 'Loess-palaeosol sequences in China and Europe: Common values and geoconservation issues', *Catena* 117: 108–18.

Veevers, J. J. (2004). 'Gondwanaland from 650–500 Ma assembly through 320 Ma merger in Pangea to 185–100 Ma breakup: Supercontinental tectonics via stratigraphy and radiometric dating', *Earth-Science Reviews* 68: 1–132.

Verner, M. (2001). *The Pyramids*, Atlantic.

Veron, A., J. P. Goiran, C. Morhange, N. Marriner and J. Y. Empereur (2006). 'Pollutant lead reveals the pre-Hellenistic occupation and ancient growth of Alexandria, Egypt', *Geophysical Research Letters* 33(6): L06409.

Wagland, S. and D. M. Gomes. (2016). 'It's time for businesses to get their hands dirty and embrace landfill mining', https://www.businessgreen.com/bg/opinion/2454124/its-time-for-businesses-to-get-their-hands-dirty-and-embrace-landfill-mining.

Wagner, M. and F.-W. Wellmer (2009). 'A Hierarchy of Natural Resources with Respect to Sustainable Development – a Basis for a Natural Resources Efficiency Indicator', *Mining, Society, and a Sustainable World*, ed. J. Richards, Springer.

Walker, J. (2006). 'What Gives Gold that Mellow Glow?', https://www.fourmilab.ch/documents/golden_glow.

Waltham, T. (2005).'The rich hill of Potosi', *Geology Today* 21(5): 187–90.

Warren, K. and A. Read. (2014). 'Landfill Mining: Goldmine or Minefield?', https://waste-management-world.com/a/landfill-mining-goldmine-or-minefield.

Watson, P. (2012). *The Great Divide: History and Human Nature in the Old World and the New*, Weidenfeld & Nicolson.

Weijers, J. W. H., S. Schouten, A. Sluijs, H. Brinkhuis and J. S. Sinninghe Damsté (2007). 'Warm arctic continents during the Palaeocene–Eocene thermal maximum', *Earth and Planetary Science Letters* 261(1): 230–8.

Wel, S. v. d. (2014). 'Religions explained', http://www.thoughtmash.net/2014/09.

Wells, N. C. (2012). *The Atmosphere and Ocean: A Physical Introduction*, 3rd edition, Wiley.

Weng, J. K. and C. Chapple (2010). 'The origin and evolution of lignin biosynthesis', *New Phytologist* 187(2): 273–85.

White, M. and N. Ashton (2003).'Lower Palaeolithic Core Technology and the Origins of the Levallois Method in North-Western Europe', *Current Anthropology* 44(4): 598–609.

White, S. (2012).'Climate Change in Global Environmental History', *A Companion to Global Environmental History*, ed. J. R. McNeill and E. S. Mauldin, Blackwell.

Wignall, P. B. (2017). *The Worst of Times: How Life on Earth Survived Eighty Million Years of Extinctions*, Princeton University Press.

Winchester, S. (2011). *Atlantic: A Vast Ocean of a Million Stories*, HarperPress.

Wing, S. L., G. J. Harrington, F. A. Smith, J. I. Bloch, D. M. Boyer and K. H. Freeman (2005). 'Transient floral change and rapid global warming at the Paleocene-Eocene boundary', *Science* 310(5750): 993–6.

Winkless, L. (2017). 'Sweating on the Underground: Why Are London's Tube Tunnels So Hot?', https://www.forbes.com/sites/lauriewinkless/2017/06/22/sweating-on-the-underground-why-are-tube-tunnels-so-hot/.

Wong, E. (2010). 'China's Money and Migrants Pour Into Tibet', https://www.nytimes.com/2010/07/25/world/asia/25tibet.html.

Woodburne, M. O., G. F. Gunnell and R. K. Stucky (2009). 'Climate directly influences Eocene mammal faunal dynamics in North America', *Proceedings of the National Academy of Sciences of the United States of America* 106(32): 13399–403.

Woodward, J. (2014). *The Ice Age: A Very Short Introduction*, Oxford University Press.

Wright, J. D. and M. F. Schaller (2013). 'Evidence for a rapid release of carbon at the Paleocene-Eocene thermal maximum', *Proceedings of the National Academy of Sciences* 110(40): 15908–13.

Wright, R. (2006). *A Short History of Progress*, Canongate.

Yong, E. (2015). 'Why Pumpkins and Squashes Aren't Extinct', https://www.nationalgeographic.com/science/phenomena/2015/11/16/why-pumpkins-and-squashes-arent-extinct/.

Zalasiewicz, J. (2012). *The Planet in a Pebble: A Journey into Earth's Deep History*, Oxford University Press.

Zalasiewicz, J., C. N. Waters and M. Williams (2014). 'Human bioturbation, and the subterranean landscape of the Anthropocene', *Anthropocene* 6: 3–9.

Zalasiewicz, J., M. Williams, R. Fortey, A. Smith, T. L. Barry, A. L. Coe, P. R. Bown, P. F. Rawson, A. Gale, P. Gibbard, F. J. Gregory, M. W. Hounslow, A. C. Kerr, P. Pearson, R. Knox, J. Powell, C. Waters, J. Marshall, M. Oates and P. Stone (2011). 'Stratigraphy of the Anthropocene', *Philosophical Transactions of the Royal*

Society A – Mathematical, Physical and Engineering Sciences 369(1938): 1036–55.

Zeebe, R. E., A. Ridgwell and J. C. Zachos (2016). 'Anthropogenic carbon release rate unprecedented during the past 66 million years', *Nature Geoscience* 9: 325–9.

Acknowledgements

The first hat tip for any large writing project should always go to the person who provided unwavering encouragement and guidance right from its very inception, and so an enormous thank you to my wonderfully supportive agent, Will Francis. Huge thanks as well to Rebecca Folland, Ellis Hazelgrove and Kirsty Gordon, also at Janklow and Nesbit in London, and PJ Mark, Michael Steger and Ian Bonaparte in the New York office. I am of course also exceedingly grateful to Stuart Williams at The Bodley Head for so enthusiastically taking this book on board for publication, and especially to Jörg Hensgen who has once again edited my manuscripts with incredible skill and thoughtfulness. Eoin Dunne helped with the figures, and the gorgeous jacket design is by Kris Potter. Thanks also to Alison Davies, Ceri Maxwell Hughes and Anna-Sophia Watts at Penguin Random House.

Many scientists and historians have also been tremendously generous with their time during the researching and writing of this book, and so thanks too (in alphabetical order) to: Christopher Beard, Davina Bristow, Alastair Culham, Steve Dutch, Chris Elvidge, Ahmed Fasih, Mike Gill, Philip Gingerich, Richard Harding, Will Hawthorne, Nicholas Klingaman, Paul Lockard, Josephine Martin, Mark Maslin, Augasta McMahon, Ted Nield, Lincoln Paine, Nicholas Rodger, Dave Rothery, Mark Sephton, James Sherwin-Smith, Ruth Siddall, Grace Steed, Phil Stevenson, Dorrik Stow, Stuart Thompson, Christiaen van Lanschot, Christopher Ware, Shoshana Weider, Chuck Williams, Scott Wing and Jan Zalasiewicz.

It has been an absolute pleasure and privilege to work with every one of you.

Figure credits

Page 20 Figure created by the author with *Mathematica* 11.0, using information from Trauth (2007) and Maslin (2014).

Pages 26–7 Figure created by the author with *Mathematica* 11.0, based on Force (2010) Figure 1, and using data on plate boundaries from Peter Bird, University of California, Los Angeles (http://peterbird.name/oldFTP/PB2002/).

Pages 34–5 Figure created by the author with *Mathematica* 11.0, using information from Woodward (2014) and Planetary Visions (http://www.planetaryvisions.com/Project.php?pid=2226).

Page 37 Figure created by the author.

Page 41 Figure created by the author with *Mathematica* 11.0, based on information in the International Chronostratigraphic Chart produced by the International Commission on Stratigraphy (http://www.stratigraphy.org/index.php/ics-chart-timescale).

Pages 50–1 Figure created by the author with *Mathematica* 11.0, using information from Metspalu (2004), Krause (2007), McNeill (2012b) Map 24.1, and Lopez (2015).

Pages 68–9 Figure created by the author with *Mathematica* 11.0, using information from Diamond (2003), Price (2009), Fuller (2014), Larson (2014).

Page 71 Figure created by the author with *Mathematica* 11.0, using river data from Natural Earth (www.naturalearthdata.com).

Page 92 Figure created by the author with *Mathematica* 11.0, using river data from Natural Earth (www.naturalearthdata.com).

Pages 102–3 Time sequence showing closure of Tethys from Stow (2010), reproduced with kind permission.

Pages 104–5 Current state of Mediterranean with mountain ranges created by the author with *Mathematica* 11.0 based on Stow (2010) Figure 29.

Pages 108–9 Figure created by the author with *Mathematica* 11.0.

Page 112–13 Figure created by the author with *Mathematica* 11.0.

Page 123 Figure created by the author with *Mathematica* 11.0. Cretaceous rock exposure distribution provided by United States Geological Survey (https://pubs.er.usgs.gov/publication/70136641), and with map reprojection help from Ahmed Fasih.

Page 151 Figure created by the author with *Mathematica* 11.0, based upon UK geological map (www.bgs.ac.uk/discoveringGeology/geologyOfBritain/makeamap/map.html) with the permission of the British Geological Survey.

Page 186 Figure created by the author with *Mathematica* 11.0 and river data from Natural Earth (www.naturalearthdata.com).

Page 188–9 Figure created by the author with *Mathematica* 11.0, using information from Bernstein (2009) Map 1, Frankopan (2016), Silk Road Encyclopedia (www.silkroadencyclopedia.com/Images2/MapSilkRoadRoutesTurkeyChina.jpg) and Travel China Guide (https://www.travelchinaguide.com/images/map/silkroad/scenery.gif).

Pages 198–9 Figure created by the author with *Mathematica* 11.0, using steppes extent from Jiao (2017) and lines of the Great Wall based on map drawn by Maximilian Dörrbecker (https://commons.wikimedia.org/wiki/File:Map_of_the_Great_Wall_of_China.jpg).

Page 221 Figure created by the author with *Mathematica* 11.0.

Page 235 Figure designed by the author and drawn by Matthew Broughton, based on Lutgens (2000) Figure 7.5, Wells (2012) Figure 6.13.

Page 243 Figure created by the author with *Mathematica* 11.0, using information from Atlas of the World (2014).

Pages 250–1 Figure created by the author with *Mathematica* 11.0, using information from Jones (2004) Figure 3.1, Bernstein (2009) Map 14, Winchester (2011) p. 319, Wells (2012) Figure 6.14.

Page 267 Figure created by the author with *Mathematica* 11.0, using information from Ulmishek (1999) Plate 3, Veevers (2004), Thomas (2013) Figure 3.2, and British Geological Society (http://earthwise.bgs.ac.uk/index.php/Regional_structure_of_the_Carboniferous,_Southern_Uplands).

Pages 270–1 Figure created by the author with *Mathematica* 11.0, using General Election 2017 data from the UK Parliament (https://researchbriefings.parliament.uk/ResearchBriefing/Summary/CBP-7979) and coal field data from the Northern Mine Research Society (www.nmrs.org.uk/mines-map/coal-mining-in-the-british-isles/), and with map reprojection help from Ahmed Fasih.

Pages 278 Figure created by the author with *Mathematica* 11.0, using information from Ulmishek (1999) Plate 5, and Veevers (2004).

Pages 282–3 Figure created by the author by digital manipulation of global image created by the NOAA-NESDIS-National Geophysical Data Center-Earth Observation Group.

Index